本书由中国科学院数学与系统科学研究院资助出版

数学 24/7

旅行中的数学

〔美〕海伦·汤普森　著

冯　雷　译

科学出版社

北　京

图字：01-2015-5625号

内 容 简 介

旅行中的数学是“数学生活”系列之一，内容涉及购买机票、制定旅行预算、飞机上行李的大小限制、不同国家货币之间的换算、租车费用及使用成本，以及不同速度单位之间的换算等数学知识，让青少年在学校学到的数学知识应用到旅行中的多个方面中，让青少年进一步了解数学在日常生活中是如何运用的。

本书适合作为中小学生的课外辅导书，也可作为中小学生的兴趣读物。

图书在版编目（CIP）数据

旅行中的数学/（美）海伦 · 汤普森（Helen Thompson）著；冯雷译.—北京:科学出版社,2018.5

（数学生活）

书名原文：Travel Math

ISBN 978-7-03-056746-8

Ⅰ.①旅… Ⅱ.①海… ②冯… Ⅲ.①数学-青少年读物 Ⅳ.①O1-49

中国版本图书馆CIP数据核字（2018）第046667号

责任编辑:胡庆家 / 责任校对:邹慧卿

责任印制:肖 兴 / 封面设计:陈 敬

科学出版社 出版

北京东黄城根北街16号

邮政编码：100717

http://www.sciencep.com

北京汇瑞嘉合文化发展有限公司 印刷

科学出版社发行 各地新华书店经销

*

2018年5月第 一 版 开本:889×1194 1/16

2018年5月第一次印刷 印张:4 1/4

字数:70 000

定价：98.00元（含2册）

（如有印装质量问题，我社负责调换）

引 言

你会如何定义数学？它也许不是你想象的那样简单。我们都知道数学和数字有关。我们常常认为它是科学，尤其是自然科学、工程和医药学的一部分，甚至是基础部分。谈及数学，大多数人会想到方程和黑板、公式和课本。

但其实数学远不止这些。例如，在公元前5世纪，古希腊雕刻家波留克列特斯曾经用数学雕刻出了“完美”的人体像。又例如，还记得列昂纳多·达·芬奇吗？他曾使用有着赏心悦目的尺寸的几何矩形——他称之为“黄金矩形”，创作出了著名的画作——蒙娜丽莎。

数学和艺术？是的！数学对包括医药和美术在内的诸多学科都至关重要。计数、计算、测量、对图形和物理运动的研究，这些都被融入到音乐与游戏、科学与建筑之中。事实上，作为一种描述我们周围世界的方式，数学形成于日常生活的需要。数学给我们提供了一种去理解真实世界的方法——继而用切实可行的途径来控制世界。

例如，当两个人合作建造一样东西时，他们肯定需要一种语言来讨论将要使用的材料和要建造的对象。但如果他们建造的过程中没有用到一个标尺，也不用任何方式告诉对方尺寸，甚至他们不能互相交流，那他们建造出来的东西会是什么样的呢？

事实上，即便没有察觉到，但我们确实每天都在使用数学。当我们购物、运动、查看时间、外出旅行、出差办事，甚至烹饪时都用到了数学。无论有没有意识到，我们在数不清的日常活动中用着数学。数学几乎每时每刻都在发生。

很多人都觉得自己讨厌数学。在我们的想象中，数学就是枯燥乏味的老教授做着无穷无尽的计算。我们会认为数学和实际生活没有关系；离开了数学课堂，在真实世界里我们再不用考虑与数学有关的事情了。

然而事实却是数学使我们生活各方面变得更好。不懂得基本的数学应用的人会遇到很多问题。例如，美联储发现，那些破产的人的负债是他们所得收入的1.5倍左右——换句话说，假设他们年收入是24000美元，那么平均负债是36000美元。懂得基本的减法，会使他们提前意识到风险从而避免破产。

作为一个成年人，无论你的职业是什么，都会或多或少地依赖于你的数学计算能力。没有数学技巧，你就无法成为科学家、护士、工程师或者计算机专家，就无法得到商学院学位，就无法成为一名服务生、一位建造师或收银员。

体育运动也需要数学。从得分到战术，都需要你理解数学——所以无论你是想在电视上看一场足球比赛，还是想在赛场上成为一流的运动员，数学技巧都会给你带来更好的体验。

还有计算机的使用。从农庄到工厂、从餐馆到理发店，如今所有的商家都至少拥有一台电脑。千兆字节、数据、电子表格、程序设计，这些都要求你对数学有一定的理解能力。当然，电脑会提供很多自动运算的数学函数，但你还得知道如何使用这些函数，你得理解电脑运行结果的含义。

这类数学是一种技能，但我们总是在需要做快速计算时才会意识到自己需要这种技能。于是，有时我们会抓耳挠腮，不知道如何将学校里学的数学应用在实际生活中。这套丛书将助你一马当先，让你提前练习数学在各种生活情境里的运用。这套丛书将会带你入门——但如果想掌握更多，你必须专心上数学课，认真完成作业，除此之外再无捷径。

但是，付出的这些努力会在之后的生活里——几乎每时每刻（24/7）——让你受益匪浅！

目　　录

Contents

1 购买机票

小李全家正在计划外出度假，他们将利用多种交通工具前往澳大利亚。在此之前，小李甚至从来没有出过国，所以他真的很期待他的首次出国旅行。

为了这次旅行，小李和他的家人花了很长时间来攒钱。小李不记得是否有过更大的旅行计划，因为他的家人一直梦想去澳大利亚。仅飞机票就要花很多钱，但全家旅行只能安排在夏天，因为这个时候小李和他的妹妹才能有一个长的假期。夏天也是飞机票价格趋向最贵的时候。

他们还没有买票，因为他们在寻找更优惠的票价。小李和他的妈妈每天在网上核查他们可能找到的最优价格。小李，妈妈、爸爸和妹妹琳达，共四个人去旅行。在下一节里，看看你能不能为小李家人找到最好的优惠票。

小李和他的妈妈在网上搜索航班，下面是他们所看到的信息：

	A航空公司	B航空公司	C航空公司	D航空公司
直达	1829.99	1818.80	N/A	1818.80
转机1次	1651.79	1805.80	1639.80	1818.50

直达航班将从他家所居住的洛杉矶机场直接抵达澳大利亚悉尼。中转一次的航班，中途有一次短暂的停泊，将增加几个小时的航行时间。

他们有两个事情要考虑：他们能负担多少钱用于支付机票？他们希望多长时间能到达澳大利亚？

1. 哪种航班更便宜，直达还是中转一次？

2. 你认为哪种航班用时短，直达还是中转一次？为什么？

小李和家人决定省钱，选择时间长的航班，因此他们查询中转一次的航班。

3. A航空公司的转机航班与C航空公司的转机航班的机票差价是多少？

A航空公司为18岁以下的年轻人提供打折机票。小李和琳达都不满18岁，所以他们的机票能便宜25%。

百分比的意思是说“在100份中所占的份数”。25%就是100份中占25份。计算飞机票价格的25%是多少，你可以通过小数点向左移动两位，把百分比转换成十进制数。那么25%就变成0.25，接下来，乘以第一个数字的十进制数：

4 1651.79美元 × 0.25 =

你已经知道折扣机票的折扣是什么，你还需要了解折扣后的票价，就是从原价减去折扣部分。

5. 折扣后票价是多少？

6. 全家机票一共需要多少钱？

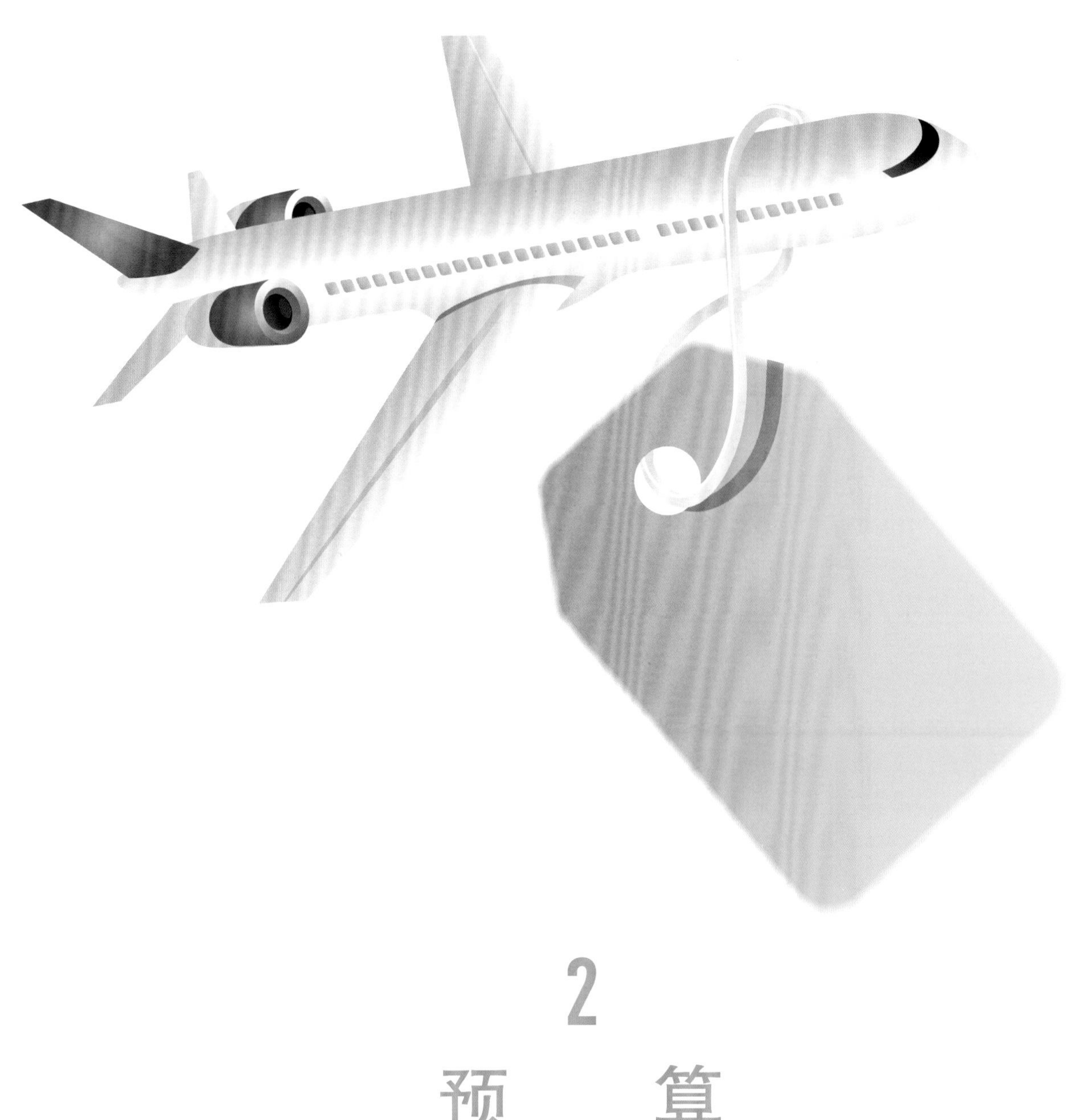

2
预　　算

现在，小李和家人已经买好了机票，他们能算出还剩多少钱可用于旅行的其他费用。他们买机票花了很多钱，所以他们的积蓄已经不多了。现在他们需要做一个预算，列出哪些方面需要花钱，可以花多少钱。

每个人都有他们想要做的事情。小李想去潜水，琳达想去海滩(尽管是7月份，但位于南半球的澳大利亚此时却是冬季)。小李的爸爸想去参观艺术博物馆，而小李的妈妈却想出去吃一次丰盛的晚餐。

他们想做的大多数事情都需要花钱，酒店住宿、购买食物、租车或乘坐公交车等。在出发之前做好预算，对他们的旅行会有很大帮助，这样他们就清楚旅行中需要花多少钱。

下面是小李家所有花钱项目的列表：

酒店房间：175美元/晚
食物：15美元/人/天
丰盛的晚餐：25美元/人
潜水：20美元/人
海滩：免费
博物馆：10美元/成人；7美元/18岁以下的孩子
交通：300美元，无论他们决定使用哪种交通工具

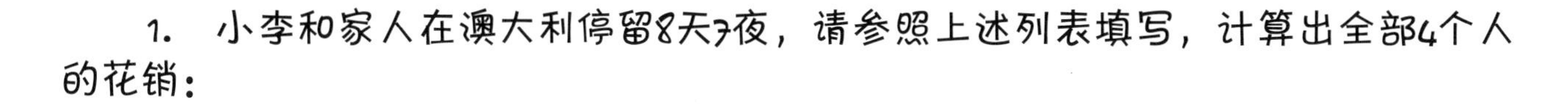

1. 小李和家人在澳大利停留8天7夜，请参照上述列表填写，计算出全部4个人的花销：

酒店房间：
食物：
丰盛的晚餐：
潜水：
海滩：
博物馆：
交通：

2. 他们将在澳大利亚旅行的总费用是多少？包括机票是多少？

在过去的几年里，小李的父母为这次旅行已积攒了7500美元。全家为这次旅行还搞了一个零钱储蓄计划。现在，储蓄罐里已有了87.75美元。

3. 他们有足够的钱满足这次旅行吗？如果没有，还需要筹备多少？

4. 距离开始旅行还有4个月时间。他们平均每个月还需要积攒多少钱才能满足这次旅行费用？

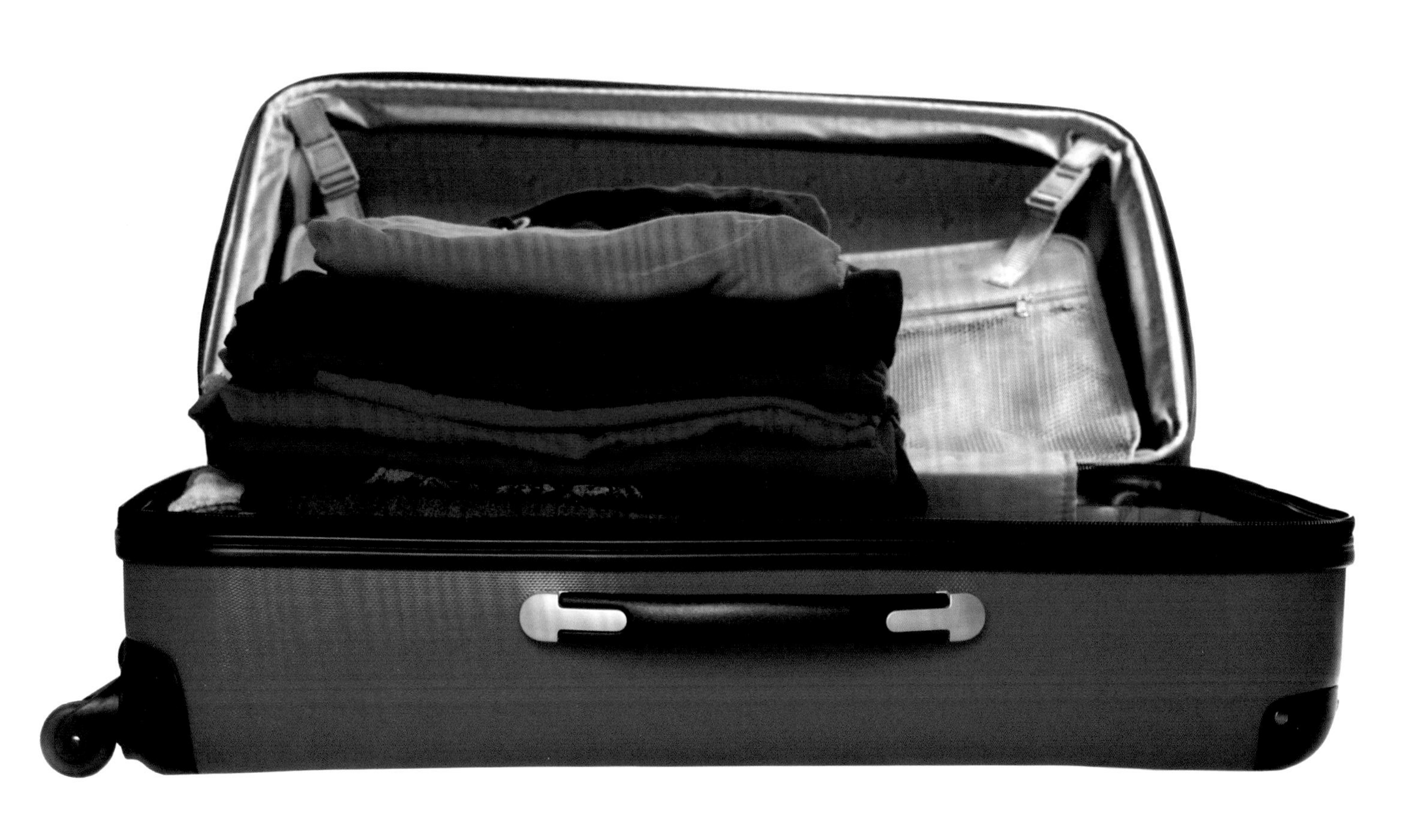

3
行李的大小

小李家的澳大利亚之行终于来了！他们明天动身，所以小李今晚必须准备好行李箱，他想带很多东西，比如衣服、游戏机、旅游方面的图书，等等。问题是，他只能装那么多。他们所乘坐的航空公司对旅客的行李大小和重量是有限制的，小李得确保行李箱不会超限，否则，要么支付额外费用，要么把行李留下。

小李用尺子测量和用秤称重来确保他的行李不会太大或太重。他的家人也必须做同样的事。下面就看看他是如何做的。

航空公司只允许每个乘客随身携带一个手提箱、一件背包(比如一个小双肩包或手提袋)及在货舱里托运一件行李。额外的行李需要另付费用。航空公司对三件行李的大小也有限制，下面是限制情况：

手提箱：
不能超过22英寸长、14英寸宽、9英寸高，所有尺寸累计(长、宽、高)加在一起不超过45英寸

背包：
长度、宽度、高度都不能超过36英寸

托运行李：
所有尺寸累计加在一起不能超过62英寸，重量必须小于50磅。

小李挑出用来托运的行李箱，经测量，尺寸是28英寸长、16英寸宽、13英寸高。

1. 箱子的尺寸、尺寸累计加在一起是多少？他能携带这件行李吗？

接下来，他称重装满东西的箱子，重52磅，这已经超出了限制。他必须把有些东西从箱子里拿出去。他看着箱子里面的东西，决定哪些东西可以留下来。

携带的物品：
额外的电子游戏，10盎司
额外的牛仔裤，1磅2盎司
登山鞋，2磅5盎司
杂志，6盎司
额外的衬衫，15盎司

他不必把所有这些东西都留在家里，只要其中一部分就可以(记住，16盎司等于1磅)。

2. 你能给出一个他能拿出去的物品组合，使他的行李箱重量少于50磅吗？

小李有2个箱子可以作为随身手提箱，一个是25英寸长、13英寸宽、8英寸高，另一种是21英寸长、14英寸宽、8英寸高。

3. 他可以携带哪一个箱子呢？

4 时区旅行

小李从洛杉矶飞到澳大利亚将跨越几个时区。洛杉矶的某个时刻，在澳大利亚却要更早些。

时间为何是不同的呢？时间只是我们用来测量太阳是如何运动的（或者说，测量地球绕太阳转动速度有多快）。一般情况下，当太阳处于天空中最高点的时候，我们称之为正午。但在地球上不同的地方，太阳将在不同的时间点位于最高点。此时此刻，或许你所在的地方正好是正午，但在地球的另一边，却正是晚上。为了保持正午的说法具有一致性，我们通过调整时间，使得当太阳高高挂在天空中时，那地方正好也是正午。

我们的办法是创建时区。每行进几百英里时间就会变换一个小时。假设你现在所处位置的时间是下午3点，那么往西几百英里的地方就是下午2点，但是东边的时间比我们早，所以往东几百英里的地方就是下午4点。当我们在南北方向上行进时，则不需要调整时间。

小李的这次旅行路途遥远，时间需要多次调整。为了避免麻烦，他必须搞清楚路途中不同地点的时间是多少。

小李和家人离开的时间是周日上午8：30。他们要飞行13个小时。

他们在新西兰的奥克兰转机，在那里停留1小时30分钟。洛杉矶和奥克兰之间的时差是+20个小时。换句话说，当洛杉矶是凌晨1点时，奥克兰是晚上9点。

然后他们从奥克兰飞到澳大利亚悉尼，飞行3小时30分钟。奥克兰和悉尼之间的时差是-2个小时。在奥克兰是晚上9点的时候，在悉尼是晚上7点。换句话说，悉尼比洛杉矶早18个小时，在洛杉矶凌晨1点的时候，在悉尼则是晚上7点。

看看他们旅行中的第一阶段，起飞时刻需增加13个小时，洛杉矶的时间是晚上9:30，那么转化为奥克兰时间：

9：30 PM 周日+20小时=5：30 PM 周一

这次旅行，需要跨越如此多的时区，竟然改变了日期！

现在你试一试完成旅行第二阶段的时间问题。

1. 在奥克兰停留一段时间后，他们什么时候飞离奥克兰？（奥克兰时间）

2. 他们到达悉尼时，奥克兰是什么时间？

现在减去2小时调整到悉尼时间。

3. 他们到达悉尼时，悉尼是什么时间？

4. 当他们到达悉尼时，洛杉矶是哪天几点？

5 平均速度

飞机起飞后，小李可以通过他面前的电视显示屏幕了解飞机的飞行过程，屏幕会显示飞机的飞行速度和飞行高度。

这架飞机飞行速度比小李以前坐过的飞机更快些！

起飞时的速度并不是很快，但飞行速度很快就提升起来，当航程快结束时，飞行速度又逐步降了下来。只要知道一些具体信息，小李就可以计算出飞机的平均速度。他肯定有足够的时间，因为他的飞行时间有13个小时！

小李收集了一些有关飞行速度的数据。每隔半小时，他记录一次飞机的飞行速度。他只能在飞机进入巡航飞行状态时，才可以解开安全带，进行记录，因此他的数据并不包括飞机起飞和着陆时的速度。

他对前6个小时的记录数据做了一个图表如下(不包括起飞阶段的半小时)：

时间	速度/MPH(英里每小时)
0小时	467
0.5小时	489
1小时	504
1.5小时	497
2小时	512
2.5小时	507
3小时	516
3.5小时	521
4小时	509
4.5小时	509
5小时	513

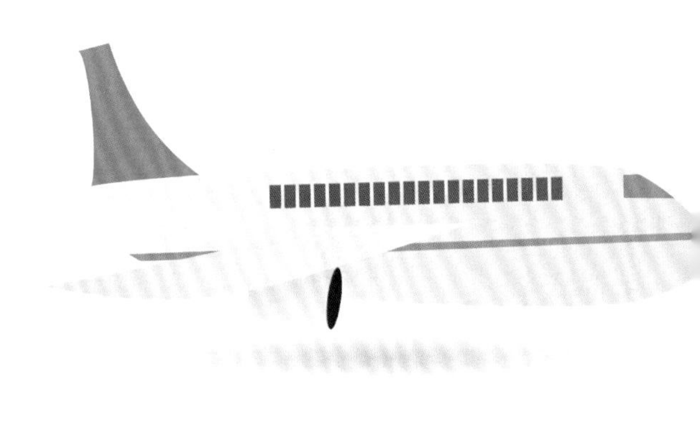

你可以通过累加上面的数值并除以个数，就能算出平均速度。

1. 小李的飞行旅程中，前6个小时的平均速度是多少？

如果知道飞行的距离或航线的长度，还可以计算出每小时飞行的平均距离，只需用距离除以飞行的时间。

洛杉矶和奥克兰(小李将在这里有一个短暂的停留)之间的距离是6520英里。

2. 用这种方法，你计算出它的平均速度是多少？它与用其他方法计算出来的速度相比，接近吗？如果不是，为什么会不同？

6

货币兑换

当小李和家人到达澳大利亚后，就需要去换钱。他们所带的现金是美元，但是在澳大利亚，人们都使用澳元，没人使用美元。他们在任何用钱的场合都需要将美元兑换成澳元。

机场有货币兑换亭，旅游者可以将任何货币兑换成澳元。小李和家人需要把一部分美元换成澳元。下面看看他们是如何做的。

到达澳大利亚后，小李发现1美元等同于0.95澳元，小李有80美元，他可以买任何他想买的东西，要想知道他能换出多少澳元，只需用美元的钱数乘以汇率0.95即可。

1. 80美元是多少澳元？

然而，小李也注意到，货币兑换亭在现金兑换时，需要收取10%的手续费。货币兑换亭先扣除10%的手续费，然后再兑换。

首先，算出小李在支付完手续费后还剩余多少美元可以兑换。

2. 小李剩余多少美元可以兑换？

然后，用上面剩余的美元钱数乘以汇率，就能得出小李能兑换到的澳元钱数。

3. 小李最后能兑换到多少澳元现金？

小李的父母带有600美元现金，在机场外，他们很容易就能找到手续费低些的换汇窗口，他们只想换能到达宾馆的费用。

4. 如果他们想兑换50美元，他们能得到多少澳元？

接着，如果他们在旅行结束后，还剩余一些钱，他们想把澳元换回美元，这一次，他们需要知道，1澳元等于1.05美元。

5. 如果他们旅行结束后，还剩100澳元，扣除手续费后，他们能换回多少美元？

7 旅行中的日程安排

小李和家人已经准备好前往这个城市，开始他们的假期！从机场到酒店，他们需要先坐火车然后换乘公共汽车。他们找到了机场的火车站，查看了列车时刻表。

他们通过研究列车时刻表，找出最佳路线。他们带了很多行李，不想在换乘期间等待太久。接下来，看看通过列车时刻表，他们是否找到了最佳路线。

他们先要乘坐火车，下车后等公交车，公交车站恰好就在他们下火车的地方。列车时刻表如下：

9:06
9:35
10:06
10:35
11:15
12:15
1:15

汽车时刻表如下：

9:07
9:37
10:07
10:37
11:07
11:37
12:07
12:37
1:07

现在时间是9:04。走到火车站台大约需要3分钟，但是小李的爸爸要去趟厕所，这需要多花几分钟的时间。

1. 他们能赶上9:06的那趟火车吗？

2. 乘坐火车需要26分钟。如果他们坐上了9:35的火车，能赶上哪趟公交车？

3. 他们等待公交车的时间会超过5分钟吗？如果会，等待时间是多长？

4. 如果需要乘坐16分钟公交车，然后还需步行4分钟才能到达酒店，那么他们最终什么时间到达？

8

租　　车

第二天早上，小李已经整装待发。虽然时差让他的生物钟有些紊乱，但是长时间疲惫的旅行昨晚让他睡得还不错。

接下来的两天，全家人享受着悉尼的风光。他们参观了爸爸想去的博物馆。吃了顿丰盛的晚餐，就像妈妈期待的那样；但是潜水和海滩可能还要等一等，因为他们马上就要离开这座城市了。

第三天的早餐后，小李的爸爸妈妈想尽快去租一辆车。比起乘坐火车而言，开车去海滩潜水会更容易一些。他们去了好几个地方，找了个租金最划算的租车公司。接下来，看看他们发现了什么。

小李的爸爸妈妈想租两天的车：今天和明天。

汽车租赁处提供给他们几个选择：每天19.95澳元租一个非常小的两门汽车，每天26.95澳元租一个中型车，或者每天35.95澳元租一辆SUV。

1. 选择不同的汽车，两天的花费分别是多少？

小李的爸爸妈妈决定租一辆中型车，这样就能装下所有人和行李。

汽车租赁公司还提供了一个租金为11澳元的GPS。小李觉得这是个好主意，因为这样他们就不会迷路了。他的父母把这个GPS也一起租了下来。

最后，汽车租赁公司还收了15澳元的保险费用。

2. 他们要付给汽车租赁公司多少钱？

事实上，汽车租赁公司还有额外的收费项目。100英里以内，小李和家人不需要支付额外费用，超出部分，每英里需额外支付0.10澳元。

今天他们想去的海滩离酒店有38英里，酒店离汽车租赁处有1.5英里。

3. 如果他们今天去海滩是开车往返的话，那么明天他们还能开多少英里而无需额外支付费用？

明天他们打算去潜水，所以他们要再次开车出城。

潜水的地方离他们的酒店只有15英里。

4. 他们最终会超过100英里的里程限制吗？如果会，他们会超过多少英里，并且要为此支付多少额外的费用？

9
一加仑汽油所行驶的里程

行驶在澳大利亚的路上，对小李来说，很多事情看起来都很奇怪。所有汽车都左侧行驶，方向盘在右边。不像在美国，所有汽车都右侧行驶，方向盘在左边。汽油的标价差别很大——看起来很贵，但小李后来意识到，澳大利亚人一定是用了不同的计量单位。

小李一家人还清楚地知道，为了到达目的地，除了租车的费用，还需支付油费。他们在把汽车还回租赁处前，还必须给汽车加满油，这样下一个客户就可以随时租车上路。

尽管小李和家人不会开车去很远的地方，但仍需要耗费一些汽油，还车时还必须加满汽油。你可以快速计算一下，他们要用多少油，要花多少钱。

下面这些概念是非常有用的：

1加仑=3.79升
1升=0.264加仑

小李已经用汽车GPS计算出他们开车到目的地的距离。第一天他们租车后将开38英里到海滩(加上他们从汽车租赁处到宾馆的1.5英里路程)。

计算开车到目的地每公里消耗了多少汽油,你既需要知道里程数，也需要知道汽油使用量。

在澳大利亚，汽油是用升来计量。在美国，以加仑来计量。车里的油表显示他们大概用了6升的汽油。

计算一升汽油行驶里程数的公式是

行驶的里程数 ÷ 用掉的汽油数

1. 他们在路途中每升汽油行驶多少英里？

澳大利亚人通常用升/百公里来表示油耗。为了得出这一数字，你需要转换上面的公式，并且换算它的单位：

用掉的汽油数 ÷ 行驶的公里数(百分制)

2. 算出小李在旅途中行驶的油耗：

6升 ÷ (10.775 × 1.61) =
1英里 = 1.61公里

小李看到的广告是：汽油的价格是137.9澳分。在澳大利亚，汽油以每升多少澳分来销售。在美国，是以每加仑多少美元来销售。你需要更多的数学运算来算出行程中的油耗。

通过单位油价乘以所消耗汽油的升数，只需向左移动两位小数就可算出以澳元为单位的钱数。

3. 他们将需要多少钱来支付那天行驶所消耗的汽油？

10
交通决策

第一天驾车出城去海滩游玩后，小李和家人返回宾馆。尽管很累，但他们还是想去吃晚餐，再多感受一下这座城市，毕竟这次来澳大利亚机会很难得。

小李非常想去看水上烟花(悉尼位于海边)。烟花燃放从晚上8点开始，现在是下午6点。他们计划吃饭大约一个小时，再留一个小时到达港口。

但是他们怎么去那里呢？他们有汽车，开车去可以快些，而且知道路线，应该不会迷路。他们也可以乘坐公共汽车，但需要花更多的时间。最后，小李计算了一下，看看哪个选择将最省钱且能按时到港口。

他们必须要考虑的是每种交通方式的成本，将开车要多少油钱和坐公交要花多少票钱进行比较。

根据GPS测算，港口离这里9英里远。他们往返将花费多少油钱？假设他们将支付137.9 澳分/升的汽油，使用上一节中的信息算出他们将消耗多少升汽油，共花多少钱。

1. 他们将使用多少升的汽油？换算成货币要花多少钱？

他们还需要支付10澳元的停车费。

2. 如果他们开车去港口的话，包括停车费和油钱，总共要付多少钱？

3. 公交车票每人每次2.50澳元，4人往返坐公交要花多少钱呢？

他们也必须考虑时间因素，在烟花开始之前，他们预留一个小时到达港口。根据GPS显示的当前交通情况，他们开车到达港口需要35分钟。小李认为他们应该考虑增加10分钟以确保能找到停车位。

与此同时，从酒店到公共汽车站要步行3分钟。公交车行程需要40分钟。下车的公交车站到港口看烟花的地方大约需要走15分钟。

4. 如果他们开车去港口要花多长时间？如果乘公共汽车呢？

5. 你认为他们应该采取什么交通方式？为什么？

11

使用地图

第二天，车上的GPS坏了。小李和家人不得不使用地图寻找潜水公司。他们开车只需要15分钟，所以去那里也不是很难。

在他们开车途中，小李可以查看地图。小李不太了解澳大利亚，因此他常看在车上备用的一个放大版澳大利亚地图。他很好奇澳大利亚到底有多大，他知道洛杉矶、加利福尼亚和美国有多大，但对澳大利亚他不确定。

地图底部有一个符号导引，注明了地图上所有符号的含义。比如，一个大点代表一个大城市。导引里还有一个距离比例尺。地图是将地球按比例缩小制成的，上面的所有图示在现实生活中都是相应存在的，只是按比例缩小了而已。下一页告诉你如何使用地图上的标尺计算距离，并比较两个地方的大小。

地图上的标尺显示1英寸相当于50公里。小李想算出横穿整个澳大利亚大陆有多远。他找到一张纸片，并在纸上以1英寸为单位做了标记，用纸片在地图上测量澳大利亚的最宽距离，算出有多少英寸。他测出整个澳大利亚大陆最宽处约80英寸。

1. 80英寸代表多少公里呢？

小李还是不太清楚那是多远，因为他不习惯用公里来描述距离。他想把距离数转换成他习惯的英里数。下面是他需要了解的知识：

1英里= 1.61公里
1公里=0.621英里

2. 横穿澳大利亚有多少英里？

在美国的中间位置，从东到西约3000英里。

3. 美国比澳大利亚宽吗？如果是，宽多少英里？

小李还有另一张地图——悉尼地图。在地图上，标尺1英寸相当于2.5公里。小李测得地图上悉尼市区最宽处约24英寸。

4. 你认为悉尼地图上显示的内容更详细还是相反？

5. 跟据李的测量，悉尼宽多少公里？是多少英里？

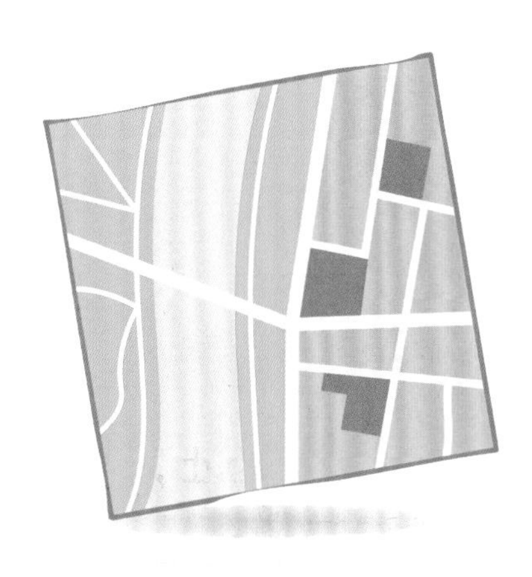

12
航海中的数学

小李和家人到达潜水服务部，他们换了泳装，拿上潜水装备，并登上船。约半个小时后，船长把他们带到一个珊瑚礁处。在这次旅行中，到目前为止，小李乘坐了飞机、火车、公交和汽车，现在又增加了一种交通工具——船！

在航行珊瑚礁途中，小李问了船长一些问题。他想知道船的航行速度有多快。船长告诉他，船的速度不是以英里每小时或公里每小时来衡量，而是节，一个仅仅用来描述船速的特殊单位。下面，他解释如何计算他们的行驶速度有多快。

1节等于1海里/每小时。不过这没用，除非你知道1海里有多远！

1海里/小时= 1.151英里/小时= 1.852公里/小时

下面是在整个珊瑚礁之旅中不同速度。请将下列图表中的节换算成英里/小时和公里/小时。

时间	节	英里/小时	公里/小时
1	9	10.36	16.67
2	13		
3	15		
4	16		
5	8		

1. 船的最快速度是多少公里/小时？

2. 在这次旅行中，船的平均速度是多少节？

船长告诉小李快艇最快时速达30节。现在因为有乘客，并且他们不希望使水下的珊瑚礁受到意外伤害，因此会开得慢很多。

3. 30节换算成英里/小时是多少？换算成公里/小时呢？

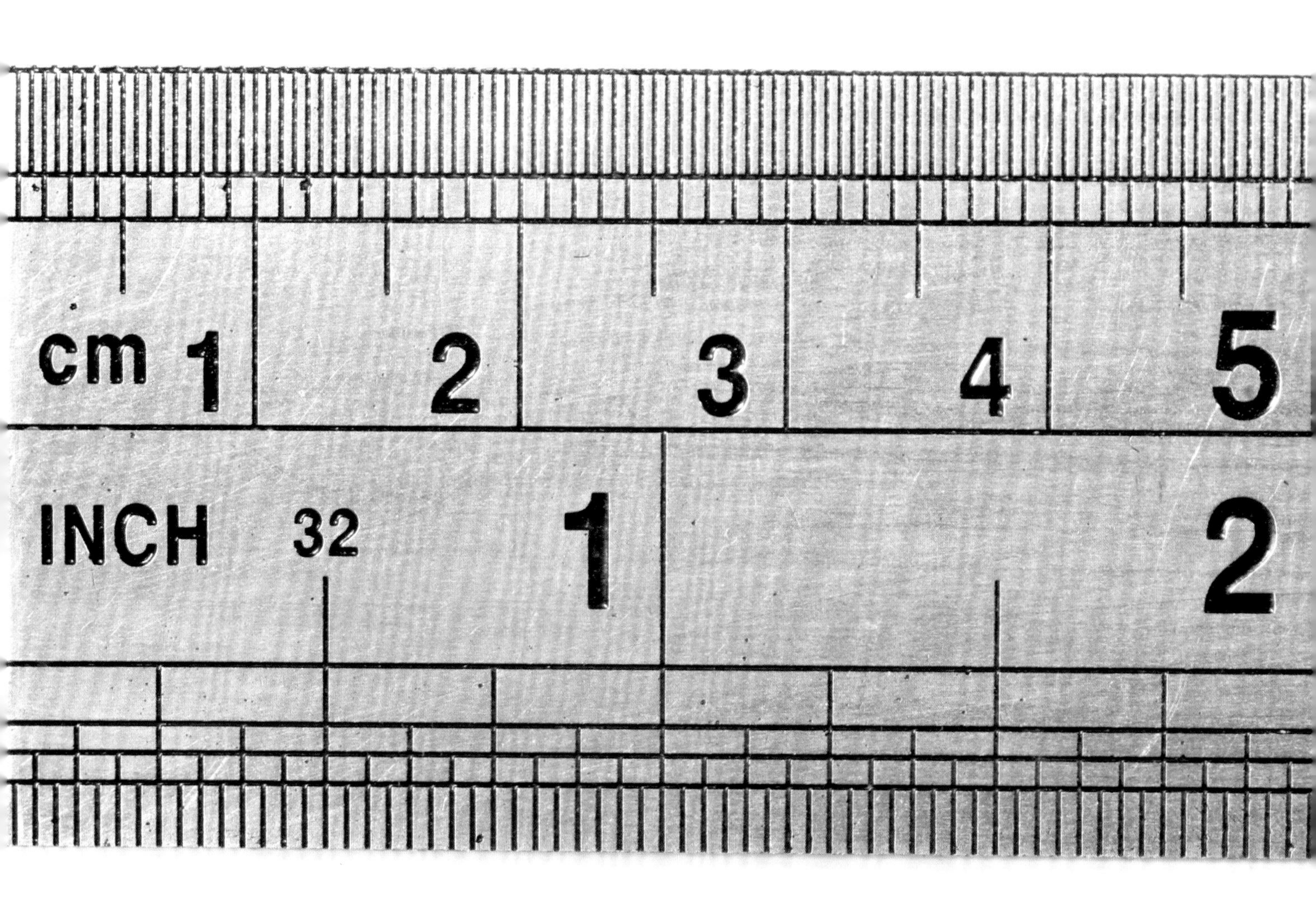

13

更多的度量转换

显然，小李已经知道了很多他不常见的度量方法。在旅行之前，他以为每个人都用英里、磅和美元作为度量单位。现在，他来到另一个国家，发现不同的人会使用不同的度量单位。例如，澳大利亚人使用不同类型的度量单位去测量距离、质量、温度，等等。他们有一套自己的度量系统。在美国，人们使用美国的系统，也称为美国标准单位。

开始小李很困惑，现在他明白了。他尝试留意所有不同的度量单位，这样他可以直接转换。下面看看他发现了什么。

小李搜集了一些度量单位及其换算，第一组换算是将其他度量转换成美国标准度量：

距离
1公里=0.621英里
1英里=1.61公里

1厘米=0.394英寸
1英寸=2.54厘米

体积
1升=0.264加仑
1加仑=3.79升

重量
1克=0.035盎司
1盎司=28.35克

1公斤=2.20磅
1磅=0.454公斤

温度
摄氏度=(华氏度-32)×5/9
华氏度=(摄氏度×9/5)+32

1. 更多关于度量换算的练习：

76公斤=
5加仑=
55华氏度=
2英寸=
52.5公里=

2. 哪个更大？把>，=，<填在下面的空白处：

30摄氏度______30华氏度
7.75厘米______4英寸
2磅______0.908公斤
89盎司______27克
1.5升______1加仑

14 旅行总成本

小李在澳大利亚的假期已经接近尾声，可他觉得好像刚刚到达那里！他和家人的假期过得很丰富也很愉快——吃东西、潜水、去博物馆、购物，等等。

他们搜集了假期中的所有收据。小李的父母想看看他们花了多少，和他们最初的预算做比较。他们是如何做的？看看下一页吧。

每个人都找出了自己的收据并且分类整理，以下是收据的汇总：

酒店房间：1220美元
食品：516澳元
丰盛的晚餐：86澳元
潜水：75澳元
博物馆：32澳元
交通：178澳元
机票：5809.30美元

所有的花销都是澳元，除了机票和住宿是他们在美国提前用美元支付的。

1. 再次填写这张图表，将所有的花费都换算成美元。
酒店房间：1220美元
食物：
丰盛的晚餐：
潜水：
博物馆：
交通：
机票：5809.30美元

他们最初的预算见第2节。

2. 他们的花费是在预算之内吗？如果是，他们花费了多少？如果不是，超出预算多少？

小李也有他自己的收据，具体如下：

T恤：23.99澳元
明信片：3.50澳元
海报：16澳元
零食：9.75澳元

他原本有80美元，在机场都兑换成了澳元。请看第6节，他换到了多少澳元。

3. 最终小李是要换更多的钱，还是控制在最初的预算内？如果没有超出预算，旅行后他还剩下多少钱？

15
小 结

在假期，小李长了很多见识，也做了很多事情。从转换度量到兑换货币、计算地图上的距离，他学到了很多数学知识。看看你是否还记得他在澳大利亚期间学到了什么和做了些什么。

1. 如果购买4张或更多的机票，航空公司可以优惠8%，那么原价630美元的机票，买4张需要多少钱？

2. 乘坐飞机时，你能随身携带一个24英寸长、13英寸宽、8英寸高的手提箱吗？为什么可以？为什么不可以？

3. 你朋友居住地的时区比你晚4小时，如果你的所在地现在的时间是10AM点，那么你朋友居住地的时间是几点？

此时给你朋友打电话合适吗？为什么？

4. 假设你去一个国家旅行，美元兑当地货币汇率是0.89。如果你想兑90美元，能兑换到多少当地货币？

如果外汇兑换处要收取12%的手续费，你最终能得到多少钱？

5. 如果租两天车的租金是35.99美元，再加上15美元的保险费，那么租一辆车需要多少钱？

6. 你的家用轿车平均油耗是32英里/加仑。如果你的家人驾车行驶了290英里，那么汽车要消耗多少汽油？

如果油箱现有11加仑汽油，是否还需将油箱装满？
如果不需要，在油耗光之前还能行驶多少英里？

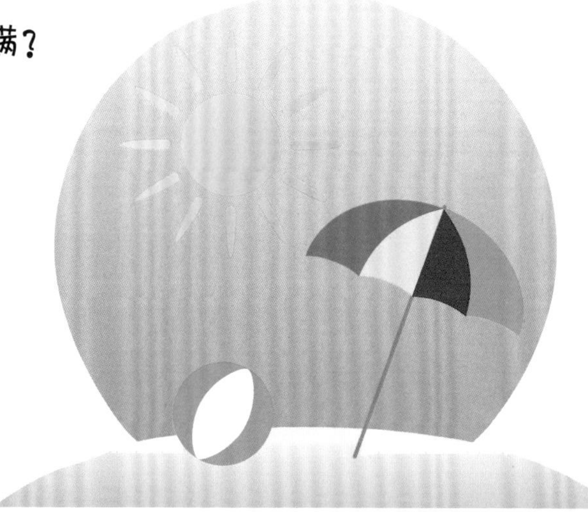

7. 你查看的地图上，1.5英寸相当于10英里，如果地图上相距15英寸，那么实际是多少英里？

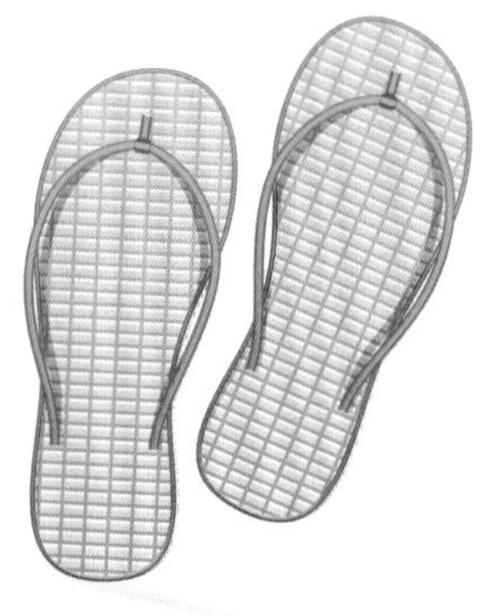

8. 78华氏度换成摄氏度是多少？

参考答案

1.

1. 中转一次
2. 直达，因为那样你就不用在中转站花费额外的时间了
3. 1651.79美元 - 1639.80美元 = 11.99美元
4. 412.95美元
5. 1651.79美元 - 412.95美元 = 1238.84美元
6. 1238.84美元 + 1238.84美元 + 1651.79美元 + 1651.79美元 = 5781.26美元

2.

1. 酒店房间：1225美元
   食物：480美元
   丰盛的晚餐：100美元
   潜水：80美元
   海滩：0美元
   博物馆：34美元
   交通：300美元
2. 2219美元；8000.26美元
3. 不，他们没有足够的钱；他们还需要412.51美元
4. 540.55/4美元 = 103.13美元

3.

1. 57英尺；能
2. 牛仔裤、电子游戏和杂志，或登山鞋和额外的衬衫，或牛仔裤和衬衫，等等
3. 第二个

4.

1. 5:30 PM + 1:30 = 7:00 PM
2. 7:00 PM + 3:30 = 10:30 PM
3. 10:30 PM - 2:00 = 8:30 PM
4. 8:30 PM - 18 小时 = 2:30 AM周一

5.

1. 504 MPH
2. 502 MPH；与用其他方法算出来的数据相比要低一些，因为在第一个平均速度中没有算上比巡航速度慢的起飞与落地的速度

6.

1. 76澳元
2. 0.1 × 80美元 = 8美元, 80美元 - 8美元 = 72美元
3. 72美元 × 0.95 = 68.4澳元
4. 50美元 × 0.1 = 5美元, 50美元 - 5美元 = 45美元, 45美元 × 0.95 = 42.75澳元
5. 100澳元 × 0.1 = 10澳元, 100澳元 - 10澳元 = 90澳元, 90澳元 ÷ 0.95 = 94.74美元

7.

1. 不能
2. 10:07那趟车 (9:35 + 26 = 10:01)
3. 会, 他们要等待6分钟
4. 10:07 + 16 + 04 = 10:27

8.

1. 39.90澳元, 53.90澳元, 71.90澳元
2. 53.90澳元 + 11澳元 + 15澳元 = 79.90澳元
3. 22.5英里 (38 + 38 + 1.5 = 77.5, 100 - 77.5 = 22.5)
4. 是的, 他们会超过7.5英里; 他们需要额外支付0.75澳元

9.

1. 77.5/6 = 12.92英里/升
2. 4.8升/100公里
3. 6升 × 137.9澳分/升 = 827.4澳分 = 8.27澳元

10.

1. 18英里/(12.92英里/升) = 1.39升; 1.39升 × 137.9澳分/升 = 191.7澳分 = 1.92澳元

2. 10澳元 + 1.92澳元 = 11.92澳元
3. 2.50美元 × 4 × 2 = 20澳元
4. 45分钟；58分钟
5. 他们应该开车，因为那样既便宜又快捷

11.

1. 4000公里左右
2. 4000公里 ×0.621 英里= 2484 英里
3. 是的，宽大约516英里
4. 更详细
5. 大约60公里，或37.26英里

12.

1. 29.63公里/时
2. 12.2节
3. 34.53英里/时；55.56公里/时

时间	节	英里/小时	公里/小时
1	9	10.36	16.67
2	13	14.96	24.08
3	15	17.27	27.78
4	16	18.42	29.63
5	8	9.21	14.82

13.

1. 76公斤 = 167.2 磅
   5加仑 = 18.95升
   55华氏度 = 12.78摄氏度
   2英尺 = 5.08 厘米
   52.5公里 = 32.6英里

2. $>$
   $<$
   $=$
   $>$
   $<$

14.

1. 酒店房间：1220美元
   食物：543.16美元
   丰盛的晚餐：90.53美元
   潜水：78.95美元
   博物馆：33.68美元
   交通：187.37美元
   机票：5809.30美元
2. 他们比预算要少花37.27美元（8000.26美元－7962.99美元＝37.27美元）
3. 他不需要换更多钱；他还剩下15.16澳元（68.40澳元－53.24澳元）

15.

1. 630美元 × 4 = 2520美元，2520 × 0.08 = 201.60美元，2520美元 － 201.60美元 = 2318.40美元
2. 不可以，因为长度超出了标准
3. 6:00 AM；不合适，因为此时你的朋友还在睡觉
4. 90 × 0.89 = 80.10澳元；70.49澳元（0.12 × 90美元 = 10.80美元，90美元 － 10.80 美元 = 79.20美元，79.20美元 × 0.89 = 70.49当地货币）
5. 86.98美元
6. 290/32 = 9.06；不需要；1.94 加仑 × 32英里/加仑 = 62.08英里，还剩62.08英里
7. 100 英里（15 英尺/1.5英尺 = 10，10 × 10英里 = 100英里）
8. 25.56

INTRODUCTION

How would you define math? It's not as easy as you might think. We know math has to do with numbers. We often think of it as a part, if not the basis, for the sciences, especially natural science, engineering, and medicine. When we think of math, most of us imagine equations and blackboards, formulas and textbooks.

But math is actually far bigger than that. Think about examples like Polykleitos, the fifth-century Greek sculptor, who used math to sculpt the "perfect" male nude. Or remember Leonardo da Vinci? He used geometry—what he called "golden rectangles," rectangles whose dimensions were visually pleasing—to create his famous *Mona Lisa*.

Math and art? Yes, exactly! Mathematics is essential to disciplines as diverse as medicine and the fine arts. Counting, calculation, measurement, and the study of shapes and the motions of physical objects: all these are woven into music and games, science and architecture. In fact, math developed out of everyday necessity, as a way to talk about the world around us. Math gives us a way to perceive the real world—and then allows us to manipulate the world in practical ways.

For example, as soon as two people come together to build something, they need a language to talk about the materials they'll be working with and the object that they would like to build. Imagine trying to build something—anything—without a ruler, without any way of telling someone else a measurement, or even without being able to communicate what the thing will look like when it's done!

The truth is: We use math every day, even when we don't realize that we are. We use it when we go shopping, when we play sports, when we look at the clock, when we travel, when we run a business, and even when we cook. Whether we realize it or not, we use it in countless other ordinary activities as well. Math is pretty much a 24/7 activity!

And yet lots of us think we hate math. We imagine math as the practice of dusty, old college professors writing out calculations endlessly. We have this idea in our heads that math has nothing to do with real life, and we tell ourselves that it's something we don't need to worry about outside of math class, out there in the real world.

But here's the reality: Math helps us do better in many areas of life. Adults who don't understand basic math applications run into lots of problems. The Federal Reserve, for example, found that people who went bankrupt had an average of one and a half times more debt than their income—in other words, if they were making $24,000 per year, they had an average debt of $36,000. There's a basic subtraction problem there that should have told them they were in trouble long before they had to file for bankruptcy!

As an adult, your career—whatever it is—will depend in part on your ability to calculate mathematically. Without math skills, you won't be able to become a scientist or a nurse, an engineer or a computer specialist. You won't be able to get a business degree—or work as a waitress, a construction worker, or at a checkout counter.

Every kind of sport requires math too. From scoring to strategy, you need to understand math—so whether you want to watch a football game on television or become a first-class athlete yourself, math skills will improve your experience.

And then there's the world of computers. All businesses today—from farmers to factories, from restaurants to hair salons—have at least one computer. Gigabytes, data, spreadsheets, and programming all require math comprehension. Sure, there are a lot of automated math functions you can use on your computer, but you need to be able to understand how to use them, and you need to be able to understand the results.

This kind of math is a skill we realize we need only when we are in a situation where we are required to do a quick calculation. Then we sometimes end up scratching our heads, not quite sure how to apply the math we learned in school to the real-life scenario. The books in this series will give you practice applying math to real-life situations, so that you can be ahead of the game. They'll get you started—but to learn more, you'll have to pay attention in math class and do your homework. There's no way around that.

But for the rest of your life—pretty much 24/7—you'll be glad you did!

1 BUYING PLANE TICKETS

Lee's family is taking a vacation. They are going all the way to Australia! Lee has never even been out of the country before, so he's really looking forward to his first international trip.

Lee and his family have been saving up money for a long time in order to go on this trip. Lee can't remember taking any big vacations, because his family has always dreamed of going to Australia. The plane tickets alone will be expensive. The family has to go during summer, because that's when Lee and his sister have a long break from school. Summer is also when plane tickets tend to cost the most.

They haven't bought the tickets yet, because they're waiting for a good deal. Lee and his mom check online every day for the best price they can find. There will be four people going—Lee, his mom, his dad, and his sister Linda. See if you can find the best deal for Lee's family on the next page.

When Lee and his mom search for flights online, this is what they see:

	Airline A	**Airline B**	**Airline C**	**Airline D**
Non-stop	1,829.99	1818.80	N/A	1818.80
One-stop	1651.79	1805.80	1639.80	1818.50

Non-stop flights will leave right from the airport right in Los Angeles, where Lee and his family live, and arrive in Sydney, Australia. Flights with one stop have a layover along the way, which adds a few hours to the trip.

Lee's family has two things to think about: how much can they afford to spend on tickets, and how fast do they want to get to Australia?

1. Which flights are cheaper, non-stop or one-stop?

2. Which flights do you think would be shorter, non-stop or one-stop? Why?

   Lee's family decides to save money and have a longer flight, so they look at one-stop flights.

3. What is the difference in prices between Airline A's one-stop flight and Airline C's one-stop flight?

Airline A offers a discount on tickets for young people under 18. Both Lee and Linda are under 18, so their plane tickets would cost 25% less.

Percents are ways of saying "parts out of 100." So 25% is 25 parts out of a hundred. To figure out what 25% of the plane ticket's price is, you can convert the percentage into a decimal number by moving the decimal point two places to the left. Then 25% becomes .25. Next, multiply the number the first number by the decimal:

4. $1651.79 x .25 =

You have just found what the discount on the ticket price was. You still need to find out what the total price with the discount is. Just subtract the discount from the original price.

5. What is the discounted ticket price?

6. How much will the plane tickets be all together, for the whole family?

2 BUDGETING

Now that Lee's family has bought the plane tickets, they can figure out how much the rest of their trip will cost. They spent a lot of money on the tickets, so they have less now in their savings. Now they will create a budget, which is a list of things they can spend money on, along with how much money they can spend.

Everyone in the family has something they want to do. Lee wants to go snorkeling. Linda wants to go to a beach (even though July will be winter in Australia because it is in the southern **hemisphere**). Lee's dad wants to go to an art museum. And Lee's mom wants to go out for a nice dinner.

Most of the things they want to do will cost money. They also have to pay for a hotel, buy food, and rent a car or take buses around the country. Making a budget before they go will help them know how much they will be spending on their vacation.

Here is a list of all the things on which Lee's family will be spending money:

hotel room: $175/night
food: $15/day per person
nice dinner: $25/person

snorkeling: $20/person
beach: free
museum: $10/adult; $7/child under 18
transportation: $300 for whichever kind of transportation they decide to use

1. Fill out the same chart, but with the costs multiplied out to account for all 4 people. They are staying in Australia for 8 days and 7 nights:

   hotel room:
   food:
   nice dinner:
   snorkeling:
   beach:
   museum:
   transportation:

2. What is the total they will spend in Australia? What is their total including airfare?

Over the past few years, Lee's parents have put aside $7500 for the trip. The family has also started a piggy bank at home to save their spare change for the trip. Right now, the piggy bank has $87.75.

3. Does the family have enough money for the whole trip? If not, how much more will they have to save to pay for the whole thing?

4. The trip is still 4 months away. How much will the family need to save each month in order to have enough for their trip?

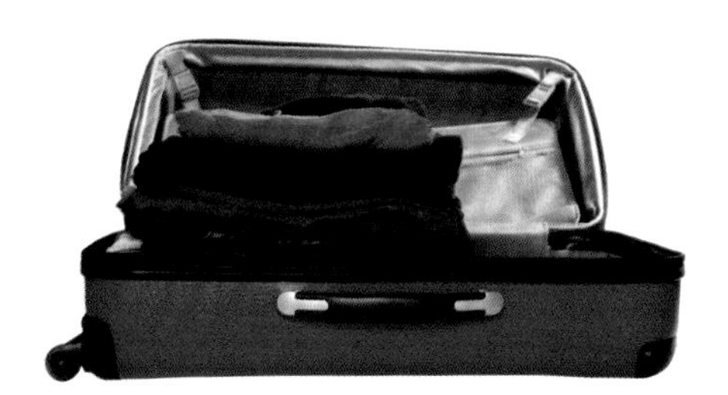

3
LUGGAGE SIZE

Lee's trip to Australia has finally arrived! They leave tomorrow, so Lee is finishing up his packing tonight. The trouble is, he can only pack so much. He wants to bring a lot with him, like clothes, games, travel books, and more. However, the airline they are flying on has limits on the size and weight of the luggage they bring. Lee has to make sure he doesn't go over the limits, or his family will have to pay extra or even leave the luggage behind!

With the help of a scale and a ruler, Lee can make sure his luggage isn't too big or too heavy. The rest of his family members have to do the same thing. See how he figures it out on the next page.

The airline only allows passengers to bring one carry-on bag, one personal item (like a small backpack or purse) and one **checked** bag that goes inside the plane in a separate compartment. Passengers have to pay for any more bags. The airline also has limits on the size of those three things. Here are the limits:

Carry-on:
Cannot be more than 22 inches long, 14 inches wide, and 9 inches tall.
All dimensions added together cannot be more than 45 inches

Personal item:
Cannot exceed 36 inches in length, width, or height

Checked bag:
All dimensions added together cannot be more than 62 inches
Must be 50 pounds or less

Lee measures the suitcase he picks out to take with him as a checked bag. It is 28 inches long, 16 inches wide, and 13 inches tall.

1. How big is the suitcase when its dimensions are added together? Can he bring it with him?

Next, he weighs the suitcase with everything in it. It weighs 52 pounds, which is more than the limit. He has to take something out. He looks in his suitcase and decides he could leave these things behind:

extra video game, 10 ounces
extra pair of jeans, 1 pound, 2 ounces
hiking boots, 2 pounds, 5 ounces
magazine, 6 ounces
extra shirt, 15 ounces

He doesn't need to leave all these things at home, just some of them. (Remember, there are 16 ounces in a pound.)

2. What is one combination of things could he take out of his suitcase to get it under 50 pounds?

Lee has a choice of 2 bags to bring as carry-ons. One is 25 inches long, 13 inches wide, and 8 inches tall. The other is 21 inches long, 14 inches wide, and 8 inches tall.

3. Which one can he bring with him?

4 TIME ZONE TRAVEL

Lee will be flying through several time zones to get to Australia from Los Angeles. When it is one time in Los Angeles, it will be earlier in Australia.

How can time be different? Time is just how we measure the movement of the sun (or really, the measure of how fast the Earth goes around the sun). In general, we say that it is noon when the sun is at its highest point in the sky. But at different places on the earth, the sun will be at its highest point at different times. Right now, it might be noon where you live. But on the opposite side of the world, it will be dark. Instead of saying that it's noon over there, we adjust time so that it will be noon when the sun is high in the sky there too.

We end up creating time zones. Every few hundred miles you travel, it gets an hour earlier.

So if it is 3:00 PM where you are now, a few hundred miles to your west it is 2:00 PM. And a few hundred miles to your east, it is 4:00 PM, because the time is ahead of you to the east. Time doesn't change when you go north or south, though.

Lee is traveling pretty far, and time will change a lot. In order not to get too confused, he'll have to figure out what time it will be at different points in his trip.

Lee and his family are leaving at 8:30 AM on Sunday morning. Their flight takes 13 hours.

They first stop in Auckland, New Zealand for a 1-hour and 30-minute layover. The time difference between Los Angeles and Auckland is +20 hours. In other words, when it is 1:00 AM in Los Angeles, it would be 9:00 PM in Auckland.

Then they fly from Auckland to Sydney, Australia. The flight is 3 hours and 30 minutes long. The time difference between Auckland and Sydney is –2 hours. When it is 9:00 PM in Auckland, it would be 7:00 PM in Sydney. Another way of thinking about it is that Sydney is 18 hours ahead of Los Angeles. When it is 1:00 AM in Los Angeles, it is 7:00 PM in Sydney.

To figure out the first part of their trip, add 13 hours to their departure time, which is 9:30 PM Los Angeles time.

Now convert Los Angeles time into Auckland time:

9:30 PM Sunday + 20 hours = 5:30 PM Monday

You end up adding so many hours that you change days!

Now try it yourself for the next part of the trip.

1. What time do they fly out of Auckland after their layover (in Auckland time)?

2. What time is it in Auckland when they land?

   Now adjust the time for Sydney by subtracting 2 hours.

3. What time is it Sydney when they finally land?

4. What time and day would it be in Los Angeles when they land in Sydney?

5 AVERAGE SPEED

After the plane takes off, Lee can watch what's going on with the plane on the TV screen in front of him. The screen shows how fast the plane is going and how high it is flying. The plane is flying a lot faster than Lee has ever gone before!

At first, the plane doesn't go very fast, but then it speeds up. By the end of the flight, it starts to slow down again. Lee can calculate the average speed of the plane if he has a little information. He definitely has enough time, since his flight is 13 hours long!

Lee collects some data on the flight's speed. Every half an hour, he writes down the speed of the plane. He only starts after the fasten-seatbelt signs are taken down, and the plane is cruising. He is looking for something called the average cruise speed, since he isn't including data on the takeoff and landing speeds.

He makes a chart like this for the first 6 hours (not including a half hour at the beginning for takeoff):

Time	Speed
0 hours	467 MPH (miles per hour)
½ hour	489 MPH
1 hour	504 MPH
1 ½ hours	497 MPH
2 hours	512 MPH
2 ½ hours	507 MPH
3 hours	516 MPH
3 ½ hours	521 MPH
4 hours	509 MPH
4 ½ hours	509 MPH
5 hours	513 MPH

You can find the average of these speeds by adding them all up and dividing by the number of speeds you added.

1. What is the average speed for the first 6 hours of Lee's flight?

If you know the distance of the flight and the length of the flight, you can also figure out the average distance traveled per hour. Just divide the distance by the number of hours flying.

The distance between Los Angeles and Auckland (where Lee will have a layover) is about 6,520 miles.

2. What is the average speed if you find it this way? Is it close to the other speed you calculated? If not, why might it be different?

6 MONEY EXCHANGE

Once Lee and his family finally land in Australia, they have to exchange their money. All they have are US dollars, but in Australia, no one uses those. Instead, they use the Australian dollar. They will have to exchange their US dollars for Australian dollars before they can spend any money.

The airport has a currency exchange booth, where travelers can convert any money they have into Australian dollars. Lee brought along some spending money he wants to exchange. His family also will exchange some of their money for Australian dollars. See how it's done on the next page.

When they get to Australia, Lee sees that 1 US dollar equals 0.95 Australian dollars (sometimes shown as A$). Lee has $80 he has saved and brought with him to spend on whatever he wants. To figure out how many Australian dollars he can get, you just need to multiply the U.S. dollars by the conversion rate, which is .95.

1. How many Australian dollars is US$80 worth?

However, he also sees that the currency exchange booth charges a 10% fee for changing money. They will take 10% of the original amount a customer wants to change.

First figure out how many U.S. dollars Lee will have to exchange after he pays the fee.

2. How many U.S. dollars will he have to exchange?

Then multiply that amount by the exchange rate to see how many Australian dollars he will have.

3. What is the final amount of Australian currency Lee will have?

Lee's parents have $600 in cash with them. They just want to exchange enough to get them to the hotel. They'll be able to find a currency-exchange business that charges a lower fee once they are out of the airport.

4. How many Australian dollars will they get if they exchange 50 U.S. dollars?

Later on, if they are left with any money at the end of the trip, they will want to change it back to U.S. dollars. This time, they need to know that 1 Australian dollar equals 1.05 U.S. dollars.

5. If they have A$100 at the end of the trip, how many U.S. dollars can they get back with the same fee?

7 TRAVEL SCHEDULES

Lee and his family are ready to head into the city and start their vacation! To leave the airport and get to their hotel, they will need to take the train and then transfer to a bus. They find the train station at the airport, and look around for the schedule.

They take a look at the schedules, and have to figure out how to line them up so the timing works. They're carrying a lot of stuff, so they don't want to end up waiting for a long time in between the train and bus. Check out the schedules on the next page and see if you can figure out the best route for them to take.

They will be taking the train, getting off, and waiting for a bus that also happens to stop at the station they got off at. The train schedule looks like this:

9:06
9:35
10:06
10:35
11:15
12:15
1:15

And the bus schedule looks like this:

9:07
9:37
10:07
10:37
11:07
11:37
12:07
12:37
1:07

Right now it is 9:04. It will take them about 3 minutes to walk to the train platform, but Linda and Lee's dad have to use the restroom first, which will take a few more minutes.

1. Will they make it to the 9:06 train?
2. The train ride lasts 26 minutes. If they take the 9:35 train, which bus can they take?
3. Will they wait more than 5 minutes for the bus? If so, how long?
4. If the bus ride to the hotel takes 16 minutes, and they have to walk another 4 minutes, at what time will they finally get to the hotel?

8 RENTING A CAR

The next morning, Lee is ready to go. His body isn't really sure what time it is, because of all the time changes. But traveling for so many hours was exhausting, and he didn't have any problem getting to sleep.

Everyone spends the next two days walking around Sydney and sightseeing. They go to the museum Lee's dad wanted to visit. They eat a nice dinner, like Lee's mom wanted. But they'll have to wait to snorkel and go to the beach, because they'll have to leave the city.

First thing after breakfast on the third day, Lee's parents want to rent a car. It will be a lot easier to drive out to the beaches and snorkeling, rather than taking the train. They go to a couple places to figure out the best car rental deal. See what they find on the next page.
Lee's parents want to rent the car for two days: the rest of today and tomorrow.

The car rental office has a few choices for them. They can get a very small car with only 2 doors for A \$19.95 a day. They could get a medium-sized car for A \$26.95 a day. Or they could get an SUV for \$35.95 a day.

1. What will the total cost be for each car for two days?

Lee's parents decide on the medium-sized car, so it fits them all and the stuff they're bringing with them.

The car rental company also offers a GPS rental for A \$11. Lee thinks that's a good idea so they don't get lost. His parents add that on to the rental.

Finally, the car rental company also charges A \$15 for insurance.

2. How much will they be paying?

It turns out that the car rental company also charges extra fees. Lee and his family are allowed to drive the rental car 100 miles for free. After that, they have to pay and additional A \$.10 a mile.

The beach they want to go to today is 38 miles from their hotel. And their hotel is 1.5 miles from the car rental office.

3. If they drive to the beach and back today, how many free miles will they have left to use tomorrow?

Tomorrow they plan on going snorkeling, which means they will be driving out of the city again. The snorkeling tour is just 15 miles from their hotel.

4. Will they end up going over their free mileage limit? If so, by how much and how much will they have to pay?

9
GAS MILEAGE

On the road in Australia, a lot of things seem strange to Lee. Everyone drives on the left side of the road, not like the United States, where everyone drives on the right side. The steering wheel is on the right side, instead of the left. The signs for gas prices also look different—they seem too expensive, but then Lee realizes that Australians must measure their gas in different **units**.

Lee's family will have to figure it out. Besides the fee for renting the car, Lee's family will have to pay for the gas to get them to their destinations. They have to return the car to the rental office with a full tank of gas, so the next customers can take it out right away.

Lee and his family aren't driving very far, but they will be using some gas, and they will have to fill the tank back up. You can do some quick calculations to determine how much gas they will need and how much they will have to spend.

These facts will be useful:

1 gallon = 3.79 litres
1 litre = .264 gallons

Lee has already used the car's GPS to find how far away his family's driving destinations are. They will drive 38 miles to get to the beach the first day they have the car. (Plus they drove 1.5

miles from the car rental office to the hotel.)

To calculate gas mileage, which is how much gas a car uses while driving a certain distance, you will need to know both the distance and the amount of gas the car used.

In Australia, gas is measured in litres. In the United States, it is measured in gallons. The gas **gauge** in the car shows that they used up about 6 litres of gas.

The equation for gas mileage is:

miles traveled ÷ gas used

1. What is the gas mileage for their trip in miles-per-litre?

Australians usually **express** mileage in litres per 100 kilometers. To get that number, you have to flip the above equation around and replace the units:

litres used ÷ kilometers traveled (in fractions of 100)

2. Solve for the mileage on Lee's trip:

$$6 \text{ litres} \div (.775 \times 1.61) =$$

1 mile=1.61kilometers

Lee sees a sign that advertises gas for 137.9. In Australia, fuel is sold in cents per liter. In the United States, it is sold in dollars per gallon. You will need to do some more math to figure out how much the gas to replace their drive will be.

Just multiply the number of litres they have used so far by the price per litre. To get the amount they will pay in dollars, move the decimal point in your answer two spaces to the left.

3. How much will they pay to replace the gas they used up that day?

10 TRANSPORTATION DECISIONS

After their first day driving the car out of the city to go to the beach, Lee's family returns to the hotel. They're pretty exhausted, but they want to go get some dinner and see more of the city. This is their only chance to visit Australia, after all!

Lee really wants to go watch some fireworks out on the water. (Sydney is on the ocean.) The fireworks start at 8:00 PM, and it is 6:00 PM right now. They figure it'll take about an hour to eat, leaving them an hour to get to the harbor.

But how should they get there? They still have the car. It would be a lot faster to drive—they know where they're going by now, so they probably won't get lost. They could also take a bus to get there, which might take more time. In the end, Lee does some calculations to see which option will be least expensive and will get them to the harbor on time.

One thing they have to think about is the cost of each. How much will gas cost versus how much will they pay for public transportation?

The harbor is 9 miles away according to the GPS. How much will the gas they are using round trip cost? Assume they will pay 137.9/litre. Use the information in the last section to figure out how much gas they will use, and how much it will cost.

1. How many litres of gas will they use? How much will it cost to replace the gas?

   They will also have to pay for parking, which will be A $10.

2. How much will they have to pay in total if they drive, including parking and gas?

3. A bus ticket will cost A $2.50 per person, per ride. How much will 4 round trip tickets on the bus cost them?

They also have to think about the time. They have an hour to get to the harbor before the fireworks start. The GPS tells them they can drive to the harbor in current traffic in 35 minutes.

Lee thinks they should add 10 minutes on to make sure they find parking.

Meanwhile, the bus stop is a 3-minute walk from the hotel. The bus ride will take 40 minutes. The bus stop at the harbor is about a 15-minute walk from where they need to be to see the fireworks.

4. How long will it take to get to the harbor if they drive? What about if they take the bus?

5. Which transportation option do you think they should take? Why?

11 USING A MAP

On the second day, the GPS in the car breaks. Lee and his family will have to use a map to get to the snorkeling company. They are only driving 15 minutes away, so it's not too hard to get there.

Lee has time to study the map as they drive. He doesn't know much about Australia, so he looks at the big Australian map that came in the car. He's curious about how big Australia is. He has a good idea about how big Los Angeles, California, and the United States are, but he's not sure about Australia.

The map has a key at the bottom, which tells what all the map symbols mean. A big dot, for example, is a big city. The key also has a distance scale. A map is a shrunken-down representation of the Earth. Everything on the map is proportional to how it is in real life, just smaller. The next page shows how you can use a map's scale to figure out distances, and compare how big two places are.

The scale on Lee's map shows that 1 inch equals 50 kilometers. Lee wants to figure out how far across the entire continent of Australia is. He finds a piece of paper and traces out 1 inch on it so he can add up how many inches it takes to get all the way across Australia. He measures almost 80 inches across the continent at its widest point.

1. How many kilometers does that represent?

Lee isn't really sure how big that is, because he's not used to measuring distance in kilometers.

He wants to convert the distance to miles, which he understands. Here's the information he needs to know:

1 mile = 1.61 kilometers
1 kilometer = .621 miles

2. How many miles across is Australia?

The United States is about 3,000 miles across from east to west in the middle of the country.

3. Is the United States wider than Australia? If so, by how many miles?

Lee also has another map. This one is of Sydney. On the Sydney map, the scale is 1 inch for every 2.5 kilometers. Lee measures about 24 inches across the widest part of Sydney, which stretches out far past the downtown area.

4. Do you think the Sydney map shows more details or fewer?

5. How many kilometers wide is Sydney, according to Lee's measurement? How many miles wide is it?

12 BOAT MATH

When Lee and his family arrive at the snorkeling business, they change into bathing suits, get their snorkeling gear, and step onto the boat. Their captain will be taking them to a coral reef about a half-hour away. Lee has ridden on planes, trains, buses, and cars so far during this trip. Now he gets to add a boat to the list!

As they sail to the reef, Lee asks the captain some questions. He wants to know how fast the boat is going. The captain tells him that boat speeds aren't measured in miles per hour or kilometers per hour. They're measured in knots, a special unit just for boat speeds. He explains how to calculate just how fast they're going, which you can do on the next page.

One knot equals 1 nautical mile per hour. That isn't helpful, though, unless you know how far a nautical mile is!

1 nautical mile = 1.151 miles per hour = 1.852 kilometers per hour

Here are the different speeds they go throughout their journey to the reef. Finish filling in the chart with miles per hour and kilometers per hour.

Time	**Knots**	**Miles per hour**	**Kilometers per hour**
1	9	10.36	16.67
2	13		
3	15		
4	16		
5	8		

1. What was the fastest the boat got in kilometers per hour?

2. What was the average speed in knots at which the boat traveled?

The captain tells Lee the speedboat can go up to 30 knots. On the snorkeling trips, they have to go more slowly because of the passengers, and because they don't want to accidentally damage any coral reefs under the water.

3. What is that speed in miles per hour? What about kilometers per hour?

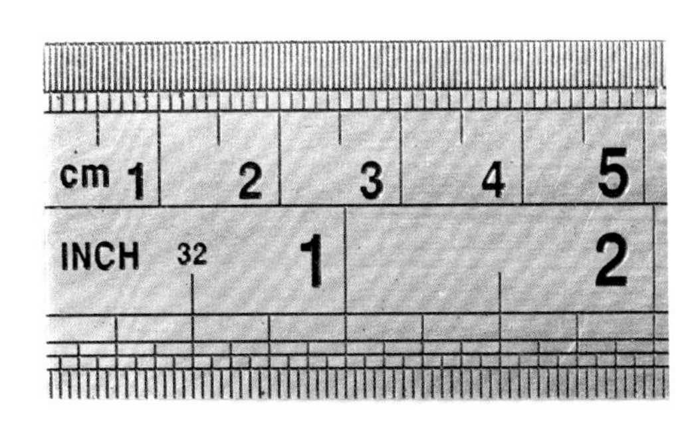

13 CONVERTING MORE MEASUREMENTS

Lee certainly has seen a lot of measurements he isn't used to. Before he traveled, he figured everyone used miles, pounds, and American dollars. Now that he's visited another country, though, he sees that different people use different measurements. For example, Australians use different types of measurements for distance, weight, temperature, and more. They use the metric system. In the United States, people use the U.S. Customary System, also called U.S. Standard Units.

At first, Lee was confused, but now he's figuring it out. He's trying to keep track of all the different measurements so he can keep them all straight. See what he has discovered on the next pages.

Here are some of the measurements and their conversions Lee has discovered. The first conversion in each pair is from metric to U.S. measurements.

DISTANCE
1 kilometer = .621 miles
1 mile = 1.61 kilometers

1 centimeter = .394 inches
1 inch = 2.54 centimeters

VOLUME
1 litre = .264 gallons
1 gallon = 3.79 litres

WEIGHT
1 gram = 0.035 ounces
1 ounce = 28.35 grams

1 kilogram = 2.20 pounds
1 pound = .454 kilograms

TEMPERATURE
Celsius = (Fahrenheit – 32) x 5/9
Fahrenheit = (Celsius x 9/5) + 32

1. Practice some more with measurement conversions:

   76 kilograms =

5 gallons =
55 degrees Fahrenheit =
2 inches =
52.5 kilometers =

2. Which is bigger? Insert >, =, or < in each blank below.

30 degrees Celsius _____ 30 degrees Fahrenheit
7.75 centimeters ______ 4 inches
2 pounds _____ .908 kilograms
89 ounces _____ 27 grams
1.5 litres _____ 1 gallon

14 TOTAL TRIP COST

Lee has reached the end of his vacation in Australia. He feels like he just got there! He and his family did a lot during there vacation time, and they had a wonderful time. They ate, snorkeled, went to museums, shopped, and more.

They collect all their receipts from their vacation. Lee's mom and dad want to see how much they spent, and compare it to the budget they made before they left. How did they do? Check the next page to find out.

Everyone finds their receipts and sorts them into categories. The following is a summary of all their receipts:

hotel room: $1220
food: A$516
nice dinner: A$86
snorkeling: A$75
museum: A$32

transportation: A$178
plane tickets: $5809.30

All of these prices are in Australian dollars, except for the plane tickets and the hotel room, which they bought ahead of time while they were still in the United States.

1. Fill out this chart again, converting all the prices to American dollars.
   hotel room: $1220
   food:
   nice dinner:
   snorkeling:
   museum:
   transportation:
   plane tickets: $5809.30

Take a look at their original estimate of how much they would spend in section 2.

2. Did they stay under budget? If so, by how much? If not, how much more money than their budget allowed did they spend?

Lee also has his own receipts for things he bought with his own money. His receipts look like this:

t-shirt: A$23.99
postcards:A$3.50
poster: A$16
snacks: A$9.75

He originally had $80, which he exchanged for Australian money at the airport. Turn to section 6 for the amount of Australian money he got.

3. Did Lee end up having to exchange more money, or did he stick to his original amount? If not, how much did he have left by the end of the trip?

15 PUTTING IT ALL TOGETHER

Lee has seen and done a lot on his vacation. He's learned a lot of math, from how to convert between measurements to exchanging money to calculating distances on maps. See if you can remember some of what he has learned and done on his trip to Australia.

1. If an airline offers a discount of 8% when you buy 4 or more tickets at once, how much would you pay for 4 tickets at $630 each?

2. Can you bring a suitcase that is 24 inches long, 13 inches wide, and 8 inches tall as a carry-on? Why or why not?

3. The time zone where your friend lives is 4 hours behind where you live. What time is it there when it is 10:00 AM where you live?

   Would that be a good time to call your friend on the phone? Why or why not?

4. You are traveling to a country where the exchange rate is .89 for the U.S. dollar. How much of the other country's money would you get if you exchanged 90 U.S. dollars?

   If the exchange business you were using charges a fee of 12%, how much money would you get?

5. How much will a car cost if you rent it at $35.99 for 2 days, and pay $15 for insurance?

6. Your family's car gets an average gas mileage of 32 miles per gallon. If your family drives 290 miles, how much gas would the car use?

Will you need to refill the gas tank if it holds 11 gallons? If not, how many more miles will it take before the gas tank is empty?

7. The scale on a map you are looking at is 1.5 inches = 10 miles. How many miles across is the area you're looking at if it is 15 inches across?

8. What is the temperature in Celsius if it is 78 degrees Fahrenheit?

Answers

1.

1. One-stop
2. Non-stop, because you don't have to add in extra time during the layover.
3. $1651.79 – $1639.80 = $11.99
4. $412.95
5. $1651.79 – $412.95 = $1238.84
6. $1238.84 + $1238.84 + $1651.79 + $1651.79 = $5781.26

2.

1. Hotel room: $1225
   Food: $480
   Nice dinner: $100
   Snorkeling: $80
   Beach: $0
   Museum: $34
   Transportation: $300
2. $2219; $8000.26
3. No, they don't have enough yet; they will need to save $412.51 more.
4. $540.55/4 = $103.13

3.

1. 57 inches; yes
2. Jeans, video game, and magazine OR hiking boots and extra shirt OR jeans and shirt, etc.

3. The second one.

4.

1. 5:30 PM + 1:30 = 7:00 PM
2. 7:00 PM + 3:30 = 10:30 PM
3. 10:30 PM – 2:00 = 8:30 PM
4. 8:30 PM – 18 hours = 2:30 AM on Monday

5.

1. 504 MPH
2. 502 MPH; it is a little lower because you didn't account for the takeoff and landing speeds in your first average, which are slower than cruise speed.

6.

1. A$76
2. .1 x $80 = $8, $80 – $8 = $72
3. $72 x .95 = A$68.4
4. $50 x .1 = $5, $50 – $5 = $45, $45 x .95 = A $42.75
5. A $100 x .1 = A $10, A $100 – A $10 = A $90, A $90 ÷ .95 = $94.74

7.

1. No
2. The 10:07 (9:35 + 26 = 10:01)
3. Yes, they will wait for 6 minutes.
4. 10:07 + :16 + :04 = 10:27

8.

1. A $39.90, A $53.90, and A $71.90
2. A $53.90 + A $11 + A $15 = A $79.90
3. 22.5 miles (38 + 38 + 1.5 = 77.5, 100 – 77.5 = 22.5)
4. Yes, by 7.5 miles; they will have to pay A $.75

9.

1. 77.5/6 = 12.92 miles/litre
2. 4.8 litres/100 kilometers

3. 6 litres x 137.9 cents = 827.4 cents = A$8.27

10.

1. 18 miles/(12.92 miles/litre) = 1.39 litres; 1.39 litres x 137.9 = 191.7 = A$1.92
2. A $10 +A $1.92 = A $11.92
3. A $2.50 x 4 x 2 = A $20
4. 45 minutes; 58 minutes
5. They should drive, because it is cheaper and faster.

11.

1. About 4,000 kilometers
2. 4,000 kilometers x .621 miles = 2,484 miles.
3. Yes, by about 516 miles.
4. More details
5. About 60 kilometers, or 37.26 miles

12.

1. 29.63 kilometers per hour
2. 12.2 knots
3. 34.53 miles per hour; 55.56 kilometers per hour

Time	Knots	Miles per hour	Kilometers per hour
1	9	10.36	16.67
2	13	14.96	24.08
3	15	17.27	27.78
4	16	18.42	29.63
5	8	9.21	14.82

13.

1. 76 kilograms = 167.2 pounds
   5 gallons = 18.95 litres
   55 degrees Fahrenheit = 12.78 degrees Celsius
   2 inches = 5.08 centimeters
   52.5 kilometers = 32.6 miles

2. >
   <
   =
   >
   <

14.

1. Hotel room: $1220
   Food: $543.16
   Nice dinner: $90.53
   Snorkeling: $78.95
   Museum: $33.68
   Transportation: $187.37
   Plane tickets: $5809.30
2. They stayed under budget by $37.27 ($8000.26 – $7962.99 = $37.27)
3. He didn't have to exchange more money; he has A$15.16 left (A$68.40 –53.24)

15.

1. $630 x 4 = $2520, $2520 x .08 = $201.60, $2520 – $201.60 = $2318.40
2. No, because the length is over the limit.
3. 6:00 AM; No, because your friend would be sleeping.
4. $90 x .89 = $80.10; $70.49 (.12 x $90 = $10.80, $90 – $10.80 = $79.20, $79.20 x .89 = $70.49)
5. $86.98
6. 290/32 = 9.06; Not yet; 1.94 gallons x 32 miles/gallon = 62.08 miles left
7. 100 miles (15 inches/1.5 inches = 10, 10 x 10 miles = 100 miles)
8. 25.56

本书由中国科学院数学与系统科学研究院资助出版

数学 24/7

时间中的数学

〔美〕詹姆斯·菲舍尔 著

王晓欢 译

科学出版社

北京

图字：01-2015-5625号

内 容 简 介

时间中的数学是“数学生活”系列之一，内容涉及时间的换算和分解、时间的将来时和过去时、24小时制和12小时制的换算、不同时区之间的换算等，同时介绍了如何让自己能准时按计划做事以及日常用的日历、秒表等，让青少年在学校学到的数学知识应用到与时间有关的多个方面，让青少年进一步了解数学在日常生活中是如何运用的。

本书适合作为中小学生的课外辅导书，也可作为中小学生的兴趣读物。

图书在版编目（CIP）数据

时间中的数学/（美）詹姆斯·菲舍尔（James Fischer）著；王晓欢译.—北京:科学出版社, 2018.5
（数学生活）
书名原文：Time Math
ISBN 978-7-03-056746-8

Ⅰ.①时… Ⅱ.①詹… ②王… Ⅲ.①数学-青少年读物 Ⅳ.①01-49

中国版本图书馆CIP数据核字（2018）第046669号

责任编辑:胡庆家 / 责任校对:邹慧卿
责任印制:肖　兴 / 封面设计:陈　敬

科学出版社 出版
北京东黄城根北街16号
邮政编码：100717
http://www.sciencep.com

北京汇瑞嘉合文化发展有限公司 印刷
科学出版社发行　各地新华书店经销
*
2018年5月第　一　版　开本:889×1194 1/16
2018年5月第一次印刷　印张:4 1/2
字数:70 000

定价：98.00元（含2册）

（如有印装质量问题，我社负责调换）

引　言

你会如何定义数学？它也许不是你想象的那样简单。我们都知道数学和数字有关。我们常常认为它是科学，尤其是自然科学、工程和医药学的一部分，甚至是基础部分。谈及数学，大多数人会想到方程和黑板、公式和课本。

但其实数学远不止这些。例如，在公元前5世纪，古希腊雕刻家波留克列特斯曾经用数学雕刻出了“完美”的人体像。又例如，还记得列昂纳多·达·芬奇吗？他曾使用有着赏心悦目的尺寸的几何矩形——他称之为“黄金矩形”，创作出了著名的画作——蒙娜丽莎。

数学和艺术？是的！数学对包括医药和美术在内的诸多学科都至关重要。计数、计算、测量、对图形和物理运动的研究，这些都被融入到音乐与游戏、科学与建筑之中。事实上，作为一种描述我们周围世界的方式，数学形成于日常生活的需要。数学给我们提供了一种去理解真实世界的方法——继而用切实可行的途径来控制世界。

例如，当两个人合作建造一样东西时，他们肯定需要一种语言来讨论将要使用的材料和要建造的对象。但如果他们建造的过程中没有用到一个标尺，也不用任何方式告诉对方尺寸，甚至他们不能互相交流，那他们建造出来的东西会是什么样的呢？

事实上，即便没有察觉到，但我们确实每天都在使用数学。当我们购物、运动、查看时间、外出旅行、出差办事，甚至烹饪时都用到了数学。无论有没有意识到，我们在数不清的日常活动中用着数学。数学几乎每时每刻都在发生。

很多人都觉得自己讨厌数学。在我们的想象中，数学就是枯燥乏味的老教授做着无穷无尽的计算。我们会认为数学和实际生活没有关系；离开了数学课堂，在真实世界里我们再不用考虑与数学有关的事情了。

然而事实却是数学使我们生活各方面变得更好。不懂得基本的数学应用的人会遇到很多问题。例如，美联储发现，那些破产的人的负债是他们所得收入的1.5倍左右——换句话说，假设他们年收入是24000美元，那么平均负债是36000美元。懂得基本的减法，会使他们提前意识到风险从而避免破产。

作为一个成年人，无论你的职业是什么，都会或多或少地依赖于你的数学计算能力。没有数学技巧，你就无法成为科学家、护士、工程师或者计算机专家，就无法得到商学院学位，就无法成为一名服务生、一位建造师或收银员。

体育运动也需要数学。从得分到战术，都需要你理解数学——所以无论你是想在电视上看一场足球比赛，还是想在赛场上成为一流的运动员，数学技巧都会给你带来更好的体验。

还有计算机的使用。从农庄到工厂、从餐馆到理发店，如今所有的商家都至少拥有一台电脑。千兆字节、数据、电子表格、程序设计，这些都要求你对数学有一定的理解能力。当然，电脑会提供很多自动运算的数学函数，但你还得知道如何使用这些函数，你得理解电脑运行结果的含义。

这类数学是一种技能，但我们总是在需要做快速计算时才会意识到自己需要这种技能。于是，有时我们会抓耳挠腮，不知道如何将学校里学的数学应用在实际生活中。这套丛书将助你一马当先，让你提前练习数学在各种生活情境里的运用。这套丛书将会带你入门——但如果想掌握更多，你必须专心上数学课，认真完成作业，除此之外再无捷径。

但是，付出的这些努力会在之后的生活里——几乎每时每刻（24/7）——让你受益匪浅！

目　录

Contents

1

时间换算

索尼娅最近一直在思考时间问题。她刚得到一部手机，可以随时看时间。每次拿出手机的时候，她都会瞄一眼时间。每天在课堂上，索尼娅也经常看教室里的时钟，算算还有多久才放学。现在快到暑假了，索尼娅已经准备好要去做运动、度假和家人、朋友们出去玩了。

索尼娅在家里的电脑上安装了一个放暑假的倒计时器。倒计时器即用天数+小时数+分钟数来倒数，也用分钟数来倒数。如果看分钟数倒数的话，那么看起来距离放暑假还远得很，但是如果用天数倒数，则暑假就近在眼前了。下面我们来看看索尼娅如何将分钟数换算成天数和小时数的。

众所周知，一小时有60分钟，一分钟有60秒。

当你想计算5小时是多少分钟，9小时23分钟是多少分钟时，你该怎么办呢？进一步地，如果换算成秒数呢？

其实，如果只在小时和分钟之间（或是只在分钟和秒钟之间）换算都是比较简单的。如果你知道小时数，只需乘以60就可以得到分钟数。

1. 请你计算一下，3小时是多少分钟？

如果是几小时零几分，那么只需将小时数换算成分钟数再加上余下的几分钟就是答案。

2. 请你换算一下，3小时56分是多少分钟？

如果想将分钟数换算成小时数，只需将分钟数除以60。

3. 请你计算一下，480分钟是多少小时？

比较棘手的情况是，用分钟数除以60不能整除的时候。例如，515分钟除以60，结果是8.58，那么后面的小数0.58小时是多少分钟呢？

当你用分钟数除以60不能整除的时候，接下来就是要计算出整数部分是多少分钟，然后用总分钟数减去这个分钟数，得到的就是后面小数表示的分钟数。例如：

$$515\text{分钟} \div 60 = 8.58\text{小时}$$

$$8\text{小时} = 480\text{分钟}$$

$$515\text{分钟} - 480\text{分钟} = 35\text{分钟}$$

那么

$$515\text{分钟} = 8\text{小时}35\text{分钟}$$

4. 171分钟可以换算成几小时又几分钟？

5. 如果距放暑假还有2天零7小时46分，那么换算成分钟是多少？提醒一下，一天是24小时。

2

时间分式

索尼娅正在和朋友们聊放学后的活动计划，她想去公园玩，但是下课后必须得先去见老师。

“去见老师只需半个小时就够了”，索尼娅说。

索尼娅的朋友杰克逊说，索尼娅要去见的老师非常喜欢聊天，“我想这次谈话至少要用三刻钟。”

索尼娅认为杰克逊说的对，建议大家四点一刻在公园里碰面，这样就能确保她有足够的时间去见老师，然后再去公园。

索尼娅和朋友们说的这些时间形式就是时间分式。就像其他东西一样，小时也可以用分式表示。接下来，我们做一些练习。

大多数人都记得住这些常用的时间分式：

一刻钟（1/4小时）=15分钟
半小时（1/2小时）=30分钟
三刻钟（3/4小时）=45分钟

1. 如果索尼娅和朋友们计划碰面的时间是四点一刻，那么具体是指什么时间？

你也可以计算出用分式表示的小时是多少分钟。只需用60除以这个分数的分母，然后再乘以分子。

我们来计算一下3/4小时是多少分钟：

3/4小时=(60 分钟 ÷ 4) × 3
3/4小时=45 分钟

2. 请你来算算，四分之二小时是多少分钟？四分之二小时还可以有什么其他的说法吗？

用上面的方法，可以计算任意一个用分式表达的小时实际是多少分钟，例如五分之一小时，三分之二小时。

3. 五分之一小时是多少分钟？

4. 三分之二小时是多少分钟？

除了小时可以用分式表达外，其他的如分钟、天、周、月、年甚至世纪也都可以用分式表达。试试计算下面用分式表达的时间：

5. 3/4周(换算成天数)=
1/2分钟(换算成秒数)=
4/5月(一个月30天，换算成天数)=
3/12年(换算成月数)=
1/4世纪(换算成年)=

3

未来时间

放学后，索尼娅和朋友们4:15在公园里碰面了。过了一会儿，杰克逊告诉大家他一会儿要去看望祖父母。索尼娅和梅根也要回家，把这学期最后一点儿作业做完。明天还有一门测验，他们都希望考得好点儿。

他们还剩下多少时间可以逗留？他们需要多少时间来学习和睡觉？索尼娅和梅根距离测验还剩多少时间？接下来我们用加法来计算时间。

索尼娅和朋友们从4:35开始讨论。一个半小时后杰克逊将离开去看望祖父母。那么杰克逊最晚几点离开才能按时到达？你可以像对待数字一样，对时间做个加法。

首先，需要将一个半小时写成时间的格式1:30，这样

$$4:35+1:30=5:65$$

由于一个小时只有60分钟，所以65分钟就要写成一个小时零五分钟，5:65就应该记为6:05。

1. 杰克逊花了大约40分钟到了祖父母家。如果杰克逊一家是6:30出发的，那么他们是几点到达祖父母家呢？

索尼娅和梅根都计划6:30之前到家吃晚饭，然后学习。梅根家就在公园附近，所以她6:30回家就可以。索尼娅家离公园有点儿远，所以她打算提前10分钟回家。那么索尼娅和梅根在公园里还能再待多长时间呢？

要回答这些问题需要做减法。你已经知道了他们要到家的时间，但你不知道在这之前还剩多少时间。

梅根6:30要离开，现在是4:35。像往常一样，我们来做减法。但请记住，1小时只有60分钟。

$$6:30-4:35=1:55$$

换句话说，梅根在离开公园前还有1小时55分钟。

想知道索尼娅在公园还能待多久，需要再做个减法。索尼娅要离开公园的时间是6:20（比梅根早10分钟），或者你也可以用梅根离开公园的时间减去10分钟来算出。

2. 如果现在是4:35，那么索尼娅离回家还剩多长时间？

4
过去时间

索尼娅回到家后吃了晚饭，然后开始为明天的数学测验做复习。想了想这一年为准备这场测验和其他数学测验所花费的时间，索尼娅想知道在数学上一共花了多少时间。她想把每天花在复习数学功课的时间加起来，当然这其中也要做一些减法。接下来我们来看看索尼娅是如何计算时间的。

首先，索尼娅做了一个清单，列出了她为准备近期的几场测验所花的时间。

第一场测验
周一：半小时
周二：1小时
周三：2小时15分钟

第二场测验
周六：45分钟
周日：2小时
周一：1小时20分钟

第三场测验
周四：4:00—5:30
周六：1:15—3:00
周日：7:00—7:30
周一：7:00—9:15

1. 索尼娅为准备每场测验花了多少时间？

2. 将上面三场测验换算成分钟是多少？

除了这三场测验，索尼娅已不记得之前的那些测验所花的时间，但是她能确定的是，这一年她一共经历了15场数学测验。

尽管索尼娅不记得为准备每场数学测验所花的时间，但是她可以算出来最近的三场测验花的平均时间。计算出平均时间后，可以用平均时间乘以15，这样就接近了她这一年为准备数学测验所花的总时间。

首先，我们来计算最近这三场测验花的平均时间。将所花的时间加起来，然后再除以3，就得到了平均时间。其实，先将这些时间换算成分钟数会比直接用几小时几分钟来计算要简单得多。

3. 索尼娅为准备每场测验所花的平均时间是多少分钟？

接下来用计算得到的平均分钟数乘以一年内的测验次数，就得到了为准备所有数学测验大概花的时间。

4. 请你来算算，索尼娅这一年为准备数学测验大约花了多少分钟？合几小时几分钟？

5
时间节点

索尼娅通常都是乘校车上学，但偶尔也会错过校车。如果发生这种情况，在街角就有公共巴士车站。她可以乘坐巴士，然后再换乘另一路巴士去学校。

索尼娅不想错过校车，因为乘坐公共巴士上学花的时间更多，而且还麻烦。然而今天早上，她还是错过了校车。更重要的是，今天是放假前的最后一天！索尼娅需要去学校跟老师和朋友们告别，并取回假期作业。

索尼娅查看了一下公共巴士时刻表。由于需要换乘，所以她得查看两条公交线路。公交车都是按点发出的，索尼娅要看明白时刻表才能不迟到。她能做到么？

现在已经7:28了，索尼娅刚刚错过了每天早上7:25左右的校车，她只好在公交车站研究一下巴士时刻表，内容如下：

1路车	2路车
7:18	7:15
7:28	7:29
7:38	7:45
7:48	7:59
7:58	8:15
8:08	8:29

1. 1路车到站时间间隔是多少分钟？

2. 2路车到站时间间隔是多少分钟？

索尼娅先坐1路公交车，要花9分钟，然后下车换乘2路公交车，还要花大约7分钟。

索尼娅已经错过了7:18和7:28的1路车，只能等7:38那趟。

3. 如果索尼娅7:38坐上了1路车，那么她大约什么时间到达2路车站？索尼娅能赶上7:45的那趟2路车么？

4. 索尼娅能赶上哪趟2路车？

学校每天8:10开始上课，如果索尼娅乘坐校车的话，大约7:50就可以到达学校，这样她还能稍微休息一下。

5. 按照上面索尼娅到2路车站的时间来看，请你计算一下，她能准时到学校么？她到达学校大约是几点？

6. 索尼娅比平时晚到学校几分钟？

6

准　时

索尼娅厌倦了错过校车和各种迟到，这让她看起来总是在疲于奔命地赶时间。这个暑假，她要培养自己守时的习惯，保证下个学期不迟到！

暑假的第一天，索尼娅有一个牙医预约。其实她并不想去看牙，但这倒是一个培养守时习惯的好机会。索尼娅的爸爸将送她去看牙医，可他也是一个经常迟到的人。索尼娅觉得只有靠自己才能保证她和爸爸准时到达。接下来，我们来看看她是如何规划时间的。

索尼娅和牙医预约的时间是下午3:30，到那里大概要花20分钟。

1. 请你算算，索尼娅应该几点出发去看牙医？

这里还有其他问题要考虑。索尼娅和爸爸要在看牙医之前填一些表格，她不知道这要花多长时间，估计在10分钟左右。在去看牙医的路上还要穿过几个施工工地，索尼娅想这大约还要多耽误15分钟。

2. 如果把这些情况都考虑在内，还要多花多长时间？他们需要几点出发？

索尼娅差点就忘了这个预约，她2:40看手机的时候才想起来。索尼娅提醒爸爸准备出发，他们出门时已经2:46。

3. 如果路上花了33分钟，那么他们能准时赴约么？如果能，那么包含填表格的时间在内，他们还剩多少时间？如果不能，会晚多久？

从现在开始，索尼娅准备做如下一张表格来规划时间。从下周开始，她要参加足球夏令营，每周一、三、五上午9点夏令营开始训练，每周二、四上午10点开始训练。从她家到夏令营训练场地要花12分钟。请你帮索尼娅把夏令营训练时间和预约牙医的时间填入下表。

事件	日期	预约时间	用时/分钟	出发时间
预约牙医	星期三	3:30		
足球夏令营				
足球夏令营				

7
闹　　钟

为了养成守时的习惯，索尼娅想应该更充分利用闹钟。上个学期，她每天都设置闹钟叫她起床上学，但也有几次，闹钟响过后，她又睡着了。

索尼娅新买的手机有闹钟功能，她想可以用这个闹钟提醒自己要守时。但是只设置一次提醒不一定会管用，所以她又另外设置了几次提醒。接下来我们看看索尼娅是如何做的。

下周，索尼娅和马克将要去游乐园，她需要设置闹钟叫她起床。马克一家大约早上8:30会来接她，所以她在这之前得做好准备。

索尼娅至少需要45分钟来梳洗打扮和吃早餐。

1. 请你计算一下，索尼娅最晚需要几点起床？

虽然索尼娅不能保证闹钟响一次就能把自己叫起来，但是闹钟有贪睡模式，闹钟响时，她可以按一下贪睡按钮，这样就可以多睡7分钟，7分钟后闹钟还会再次响起。

请你计算一下，索尼娅按了贪睡按钮后，会多睡多长时间？那么她几点起床呢？

2. 按0次=0分钟，7:45
   按1次=
   按2次=
   按3次=
   按4次=

索尼娅想她或许会按两次贪睡按钮。

3. 如果索尼娅按两次贪睡按钮的话，那么她需要将闹钟设置成几点？

在行程当天，索尼娅实际上是按了三次贪睡按钮，以至于她准备得很匆忙。

4. 如果索尼娅按照按两次贪睡按钮的时间来设置的闹钟，那么她实际上是几点起床的？她还剩多长时间来准备？

8

秒　表

在游乐园里，索尼娅和马克玩了云霄飞车等许多项目，吃了披萨，玩得非常开心。

然后他们又玩了一会儿卡丁车。游乐场安装了一套卡丁车系统让玩家用来比赛。每天，游乐场都更新最快纪录并及时公布出来，看看哪个玩家能打破当天这个纪录。索尼娅和马克在去玩之前，先看看别人玩的情况。

过了一会儿，他们发现游乐场公布的比赛用时与实际用时有出入。索尼娅用手机中的秒表来记录他们实际完成赛程需要的时间。当他们算清楚这其中的差异后，就准备下场去比试并争取赢得比赛。

赛手需要绕赛道跑5圈。目前榜单前几名的用时如下：

3分3秒
3分17秒
3分24秒
3分39秒
3分41秒

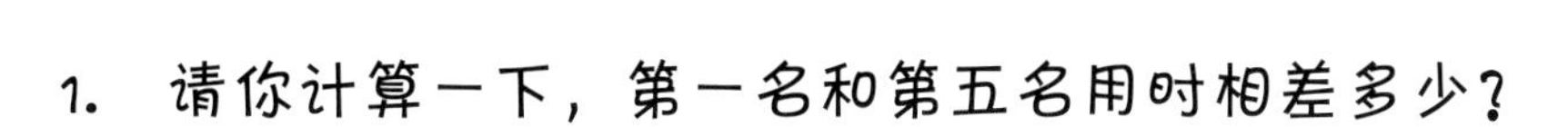

1. 请你计算一下，第一名和第五名用时相差多少？

索尼娅和马克用自己的秒表记录了比赛用时，发现游乐场公布的比赛用时是不准确的。游乐场公布的第一名赛手完成赛程的用时是3分3秒，而索尼娅和马克记录的时间是4分53秒。

2. 请你计算一下，游乐场的计时器慢了多少？

为了确认判断，他们又测试一次。这一场游乐场公布的本场冠军用时是3分38秒，而他们记录的时间是5分28秒。

3. 请你计算一下，游乐场计时器的误差两次测试结果是否一样？如果不一样，那么这一次误差是多少？

马克准备下场比赛，看看自己是否能打破当天的纪录。索尼娅答应会站在赛道旁一直看着娱乐场计时器，当要接近最快纪录时，她会大喊提醒马克破纪录的时间所剩不多了。

当马克还剩最后一圈时，索尼娅大喊告诉他娱乐场计时器显示用时是2分30秒。

4. 请你计算一下，马克每圈平均用时是多少？

5. 如果马克最后一圈用时是前面几圈的平均用时，那么他能否打破当天的纪录呢？他完成比赛的用时是多少？如果马克不能打破纪录，那么距最快记录相差多少？

9

日历中的数学

索尼娅已经慢慢养成守时的习惯了，但她却常常会忘事，然后在最后时刻才想起来，这样又会迟到。就像上周她朋友米根的生日派对，索尼娅就忘记了，直到一个朋友给她打电话问她是否去参加派对时，索尼娅才记起这件事，可此时距派对开始只剩半小时了。

索尼娅为了解决忘记重要事情的烦恼，运用数学知识，用日历制作了一个日程计划表，提醒自己不要忘记约定的重要事情。

索尼娅要记住的下一个重要事情是，7月31日是她爸爸的生日。她必须记住给爸爸准备生日贺卡和礼物。

1. 今天是7月26日，距索尼娅爸爸生日还有几天？

在日历上距爸爸生日还剩两天的那一日，索尼娅做了个标记，以便提醒自己记起这件事。

2. 距日历上有提醒标记的那天还剩多少天？

索尼娅也想好好充分利用这个暑假，她不确定暑期还剩多少天，而用日历来计算就容易多了。

索尼娅可以用日历数天数的方法来计算距暑期结束还剩多少天，但是这种数天数的方法很费功夫，而且很容易数错。

她也可以用数周数的方法来计算，这个方法要比数天数容易一些，她只需将周数乘以7，这样就能计算出暑期剩余的天数。

3. 如果距索尼娅返校还有5周零5天，那么换算成天数是多少天呢？

索尼娅也可以用每月的天数来计算剩余的天数。今天是7月26日，学校的开学日期是9月4日。

如果用这个方法来计算，需要先了解每个月有多少天。有的月份有30天，有的月份有31天，有的月份有28天。

28天：二月
30天：四月、六月、九月、十一月
31天：一月、三月、五月、七月、八月、十月、十二月

4. 请你使用上面的数据来算算距索尼娅开学还有多少天？

10
时 间 轴

索尼娅的妈妈正在整理家族的历史。她通过整理老照片和过去的信件来理顺家庭成员之间的相互关系。索尼娅觉得她的家族史很有趣，所以也经常帮忙整理。因为有太多的文件和数据，她发现要想将过去的历史全部追溯出来很难。

索尼娅和妈妈将家族的这些历史材料按照时间轴来整理。每次她们拿到一个新的信息，比如谁谁出生了，就补充在时间轴中。整理完后，就得到了一个以家族重大事项为线索的时间轴。接下来我们来看看这条时间轴，并帮她们补充信息。

下面是索尼娅和妈妈收集到的1900年到1960年间的重大事项：

1908年9月30日，曾祖母出生
1912年3月4日，曾祖父出生
1927年，曾祖父母来美国
1941年，祖母出生
1953年，曾祖父去世
1958年，母系家族来美国
1960年，祖父开始做生意

时间轴从左至右表示时间由远即近。按照时间顺序，索尼娅使用年月日对时间轴进行标注。

将上面得到的索尼娅家族信息完整地补充在下面时间轴中，上面的信息是她们相继发现记下的，并没有按照时间顺序排列。信息框可以在时间轴上面，也可以在下面。

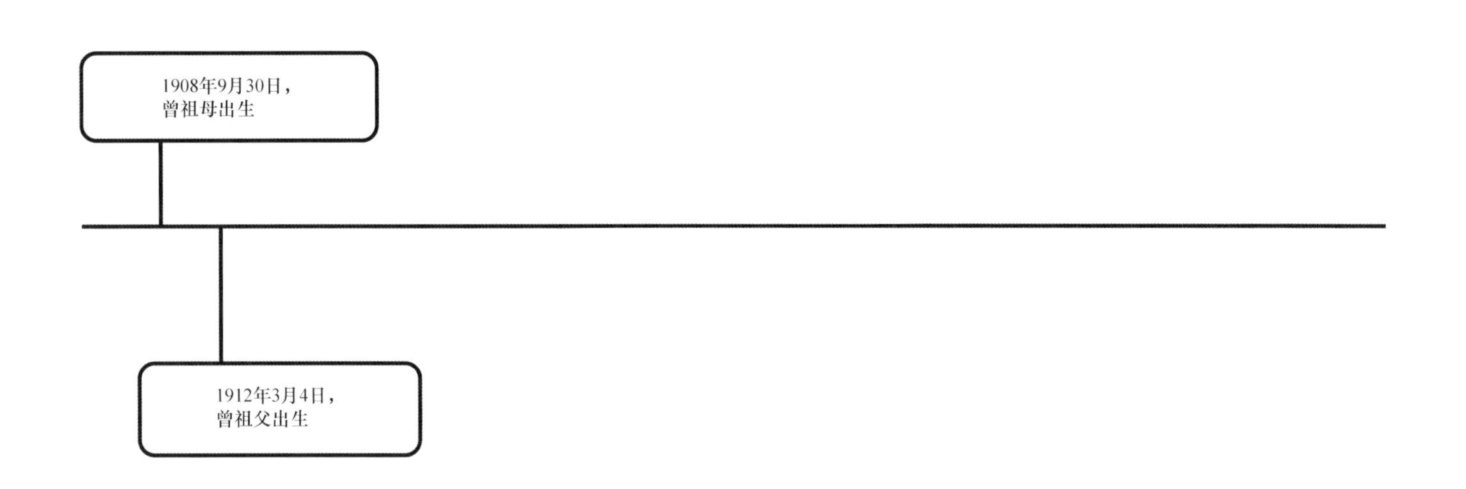

11

24小时制时间

一天下午，索尼娅去朋友娜塔莉亚家，她们玩了一会儿电脑游戏。偶然间索尼娅注意到电脑显示的时间16:34，“这是什么意思？”索尼娅指着这个时间问道。

娜塔莉亚解释说电脑上显示的都是24小时制时间（在美国，只有军队才使用24小时制时间，其他人都不这么用）。娜塔莉亚的父母来自俄罗斯，俄罗斯人习惯使用24小时制时间，这与北美地区人们的习惯不太一样。由于他们习惯了24小时制时间，所以将电脑时间设置成了24小时制的形式。

索尼娅看了看自己的手表，现在是4:34。她觉得可以搞明白这个时间与24小时制时间之间的关系。你能么？

在部分地区，人们使用12小时制，那么一天中会有两个同样的时刻。例如，8:04既代表早上8:04，也可以代表晚上8:04；午夜12点和中午12点都是12小时制时间的0点。我们一般用AM来代表上午时间，用PM代表下午时间。

在24小时制时间里，一天中的时刻是不会重复的。事实上，每一小时都有自己专属的数字，用1到24来表示。中午12:00过后就是13:00，这样就不必用AM和PM来区分上午、下午时间了。

从凌晨1:00到中午12:59在12小时制和24小时制看起来是一样的。上午11:30，用12小时制表示就是11:30AM，用24小时制表示是11:30。在这种情况下，个位时间用24小时制时间表示时前面要添上0，例如，早上2:00AM，用24小时制表示就是02:00。

1. 早上3:54 AM用24小时制时间如何表示？

中午12:59 AM和下午1:00 PM用24小时制时间表示有何不同？

将24小时制时间转换为12小时制时间的方法，是用24小时制时间减去12:00，得到的就是12小时制时间，例如：

16:34 - 12:00 = 4:34

这两种时制表示分钟是一样的，只是在表示小时的时候会有不同。

2. 24小时制19:20转换成12小时制是几点？

3. 24小时制22:45转换成12小时制是几点？

请你猜猜12小时制的凌晨12:00用24小时制如何表示？人们用凌晨12:00代表新一天的开始，用24小时制表示就是00:00。

4. 请将下面两种时制进行转换，请注意区分AM和PM。

6:00 PM = ________

________ = 05:30

________ = 23:18

12:09 PM = ________

________ = 00:20

12 罗马数字

索尼娅在娜塔莉亚家注意到一个时钟的数字很特别。钟面上的数字不是1到12，而是像I和X这样的字母。索尼娅又去请教娜塔莉亚如何从这个时钟上看时间。

娜塔莉亚说这个时钟钟面的数字是用罗马数字表示的。很久以前，罗马帝国都是使用罗马数字的。如今，人们更多是使用1，2，3，…罗马数字现在仍在使用的地方之一就是时钟。最近这几年大型的体育赛事，如超级碗和奥林匹克运动会也都是用罗马数字。因此学习罗马数字还是十分必要的。

既然你已经知道钟面上的数字是1到12，那么钟面上的罗马数字就比较容易认了。

下面是一些常见的罗马数字：

I = 1
V = 5
X = 10
L = 50
C = 100
M = 1000

对钟表而言，你只需认识I，V和X这三个罗马数字就可以了。

你可以将两个字母组合在一起来表示数字。两个I或II可以表示2，三个I或III可以表示3，但是4有点不同，我们通常不会写成IIII，而是用IV来表示，意思是比5少1。9也一样，不过是用X代替V。

1到6用罗马数字表示就是：I，II，III，IV，V，VI。

1. 请你来想想钟面上7到12用罗马数字如何表示呢？

娜塔莉亚刚好有一个用罗马数字表示的24小时制时钟，下面就是这个时钟的钟面。

2. 请你将钟面上的其他罗马数字补充完整。

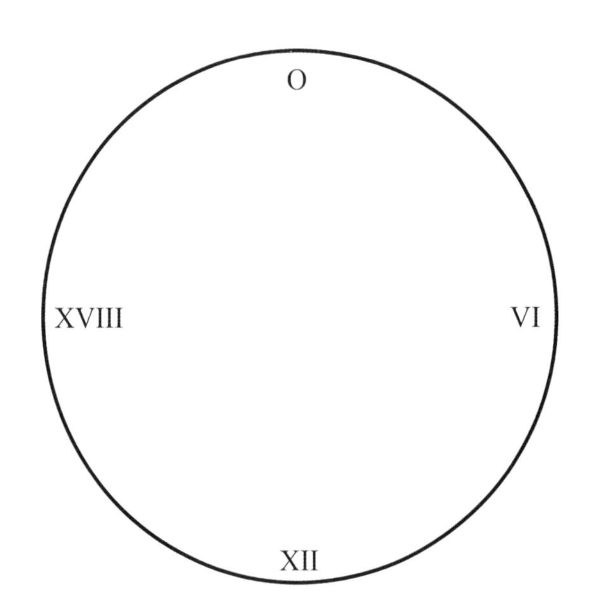

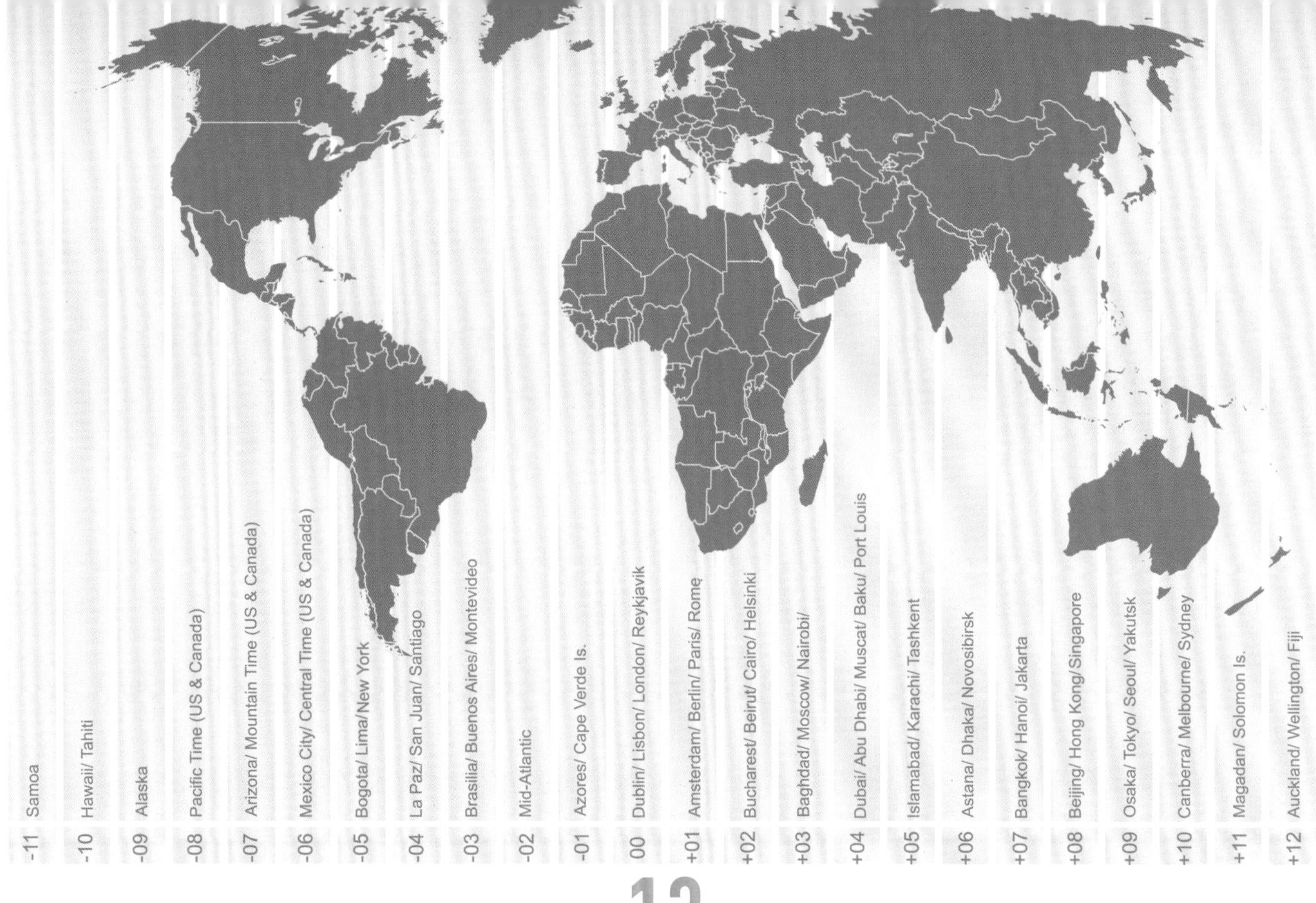

13

时　区

索尼娅的朋友约翰由于母亲在加利福尼亚谋到新职位，所以这个暑假全家将要搬到加利福尼亚。索尼娅很不舍得朋友的离开，约定好一定要保持联系，问题是加利福尼亚实在太远了。索尼娅家在东海岸，距离加利福尼亚有几千公里，两个地方不在同一时区。加利福尼亚的时间要比东海岸早一些。

时间为什么会不同呢？时间是我们描述太阳运动的工具，更确切地说，时间是描述地球围绕太阳公转的工具。一般来说，我们将太阳在空中最高点的时刻定义为正午。但是在不同的地域，太阳在最高点的时刻是不同的。例如，你现在居住的地方是正午，而在地球的另一端则是黑夜。所以为了保持正午的说法具有一致性，我们调整时间使得太阳挂在天空最高点时，那地方正好也是正午。

于是我们有了时区的概念。假如你从地球东边走到西边，每走几百里，时间就早一个小时。比如说，现在你所在的地方是下午3:00，那么在往西几百里远的地方，时间是早一个小时的。同样地，在往东几百里远的地方，时间又是晚一小时的。如果你只顺着地球南北方向走，时间是不变的。

索尼娅在与约翰联系的时候也要考虑到时差这个问题。她不想在约翰睡觉的时候给他打电话，当然她也不想睡觉的时候接到约翰的电话。索尼娅要算出给约翰打电话的最佳时间。

东海岸所在的时区是东部时区，加利福尼亚所在的时区是太平洋时区，两个时区相差3个小时。

当索尼娅所在东部时区是下午4:00时，加利福尼亚时间要早3个小时。如果你想计算出较早时区的时间，只需做个减法即可：

4:00 PM - 3:00 = 1:00 PM

1. 如果索尼娅在的东部时区时间是上午10:00，那么加利福尼亚是什么时间？这时候给约翰打电话合适么？

约翰要给住在东海岸的家人和朋友打电话，他也想知道如何计算出东部时间。

当约翰搬去的地方是下午4:00时，以前住的地方要比这个时间晚3个小时。要想计算出较晚时区的时间，只需做个加法即可：

4:00 PM + 3:00 = 7:00 PM

2. 当约翰新家的时间是晚上11:00时，索尼娅家是什么时间？约翰这个时间给索尼娅打电话合适么？为什么？

东西方向距离越远，时差就越大。比如说，娜塔莉亚的祖父母住在莫斯科郊区，娜塔莉亚和祖父母住的地方时区相差非常大，祖父母家时区比娜塔莉亚所在的时区早8个小时。

3. 如果现在娜塔莉亚的时区是下午5:00，那么她祖父母所在时区是几点？

4. 此时，娜塔莉亚和祖父母是在同一天里么？为什么？

14
穿越时区

有一天，索尼娅想去加利福尼亚拜访约翰。她还没去过加利福尼亚，而到那里度假，可以顺便去看望约翰。

去加利福尼亚旅行会涉及时区的改变。与打电话给其他时区的人不同的是，索尼娅要亲身穿越几个时区。

当人在穿越几个时区时，很容易把时间弄混，打乱作息时间。但是只要你利用数学算一算就发现这也不难解决。接下来我们看看如何解决这个问题。

你已经知道了加利福尼亚所在的时区比索尼娅在的时区早3个小时，如果索尼娅到了加利福尼亚，那么她白天的时间就会延长。

下面是索尼娅去加利福尼亚的航班信息：

出发时间：9:35AM
航程时长：4小时20分

1. 如果不考虑时区的概念，那么她到达加利福尼亚是什么时间？

然而，加利福尼亚时区比东部时区早3个小时，索尼娅到达的时间减去3小时就是当地的时间。

2. 索尼娅到加利福尼亚后，当地时间是几点？

还有一个办法来计算索尼娅到加利福尼亚后的当地时间。先计算出旅程所需时间和时差的差值，然后将差值加上索尼娅出发时的时间，就得到了她到加利福尼亚的当地时间。

3. 4小时20分－3小时=1小时20分

1:20＋9:35＝

索尼娅在去加利福尼亚的途中要转机，她要在比东部时间晚一小时的芝加哥中转。

如果索尼娅到达芝加哥的当地时间是10:10，那么她这一段飞行时长是多少？

将芝加哥时间转换为东部时间，然后两个时间相减就得到了答案。

4. 飞行时长是多久？

15
小　　结

这个暑假索尼娅学习到了很多关于时间的知识。她知道了如何守时，认识了罗马数字，懂得了24小时制时间。接下来我们看看你是否还记得这些知识。

1. 请你来换算一下，3天零7小时16分是多少分钟？

2. 请你来计算一下，三分之一小时是多少分钟？

那么，一个星期的四分之一是多少天？

3. 如果你要去乘坐公交车，下一趟车是3:07到站，再下一趟车是3:29到站，现在时间是2:58，从家走到公交车站需要11分钟。

那么你能否赶上下一趟公交车？

如果赶不上下一趟车，那么距再下一趟车还要等多久？

4. 假设你将闹钟提醒设置为早上7:15，闹钟的贪睡功能是每6分钟响一次，一共按了两次贪睡功能键。

起床后洗漱出门要用去35分钟，你能否在8:05之前到达学校？

5. 今天是1月7日，3月14日学校要举办一场校园音乐会，那么从今天算起距校园音乐会还有多少天？

6. 早上7:45AM用24小时制时间如何表示？

下午7:45PM用24小时制时间又如何表示？

7. 12用罗马数字如何表示？

8. 假设你有一个朋友住在东部，比你早两个时区。

如果现在你朋友所在的时区时间是晚上9:00PM，那么你所在的时区时间是几点呢？

参考答案

1.

1. $3\times 60=180$分钟
2. $180+56=236$分钟
3. $480/60=8$小时
4. 2小时51分钟
5. $(24\times 2\times 60)+(7\times 60)+46=3346$分钟

2.

1. 4:15
2. 30分钟；半小时
3. 12分钟
4. 40分钟
5. 5 1/4天
   30秒
   24天
   7月
   25年

3.

1. 6:30 + 0:40 = 7:10
2. 1小时45分钟

4.

1. 测试1 = 3小时45分钟；测试2 = 4小时 5分钟；测试3 = 6 小时
2. 225分钟，245分钟，360分钟
3. 276.67分钟
4. 4150分钟；69小时又10分钟

5.

1. 10
2. 14分种和16分钟交替
3. 7:47; 不能
4. 7:59到站的车
5. 能,她将于8:06到达学校
6. 16分钟

6.

1. 3:10
2. 多花25分钟; 2:45
3. 是,他们能准时到，并且还能比预约时间提前11分钟

事件	日期	预约时间	用时/分钟	出发时间
预约牙医	星期三	3:30	45	2:45
足球夏令营	星期一，星期三，星期五	9:00	12	8:48
足球夏令营	星期二，星期四	10:00	12	9:48

7.

1. 7:45
2. 7分钟, 7:52
   14分钟, 7:59
   21分钟, 8:06
   28分钟, 8:13
3. 7:31
4. 7:31 + 0:21 = 7:52; 8:30 - 7:52 = 38分钟去准备

8.

1. 35秒
2. 4:53 - 3:03 = 1分钟50秒
3. 一样，依然是1分钟50秒
4. 37.5秒
5. 不能；3分钟7.5秒(150秒 + 37.5秒= 187.5秒 = 3分钟 7.5秒)；他距最快记录相差4.5秒

9.

1. 5天
2. 3天
3. 40天
4. 40天

10.

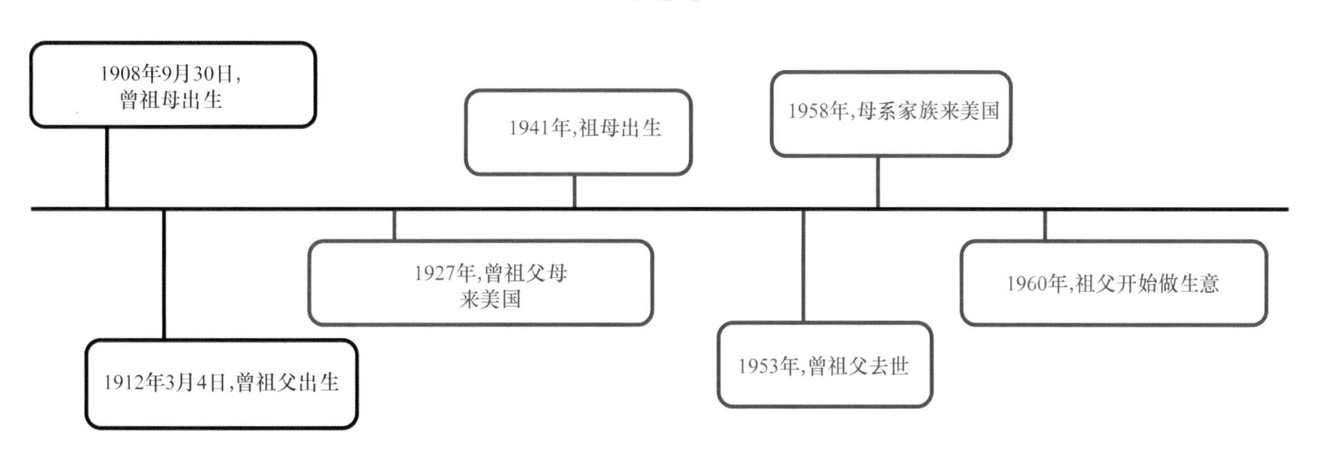

11.

1. 03:54
2. 7:20 PM
3. 10:45PM

4. 18:00
   5:30 AM
   11:18 PM
   12:09
   12:20 AM

12.

1. VII, VIII, IX, X, XI, XII
2. I, II, III, IV, V, VII, VIII, IX, X, XI, XIII, XIV, XV, XVI, XVII, XIX, XX, XXI, XXII, XXIII

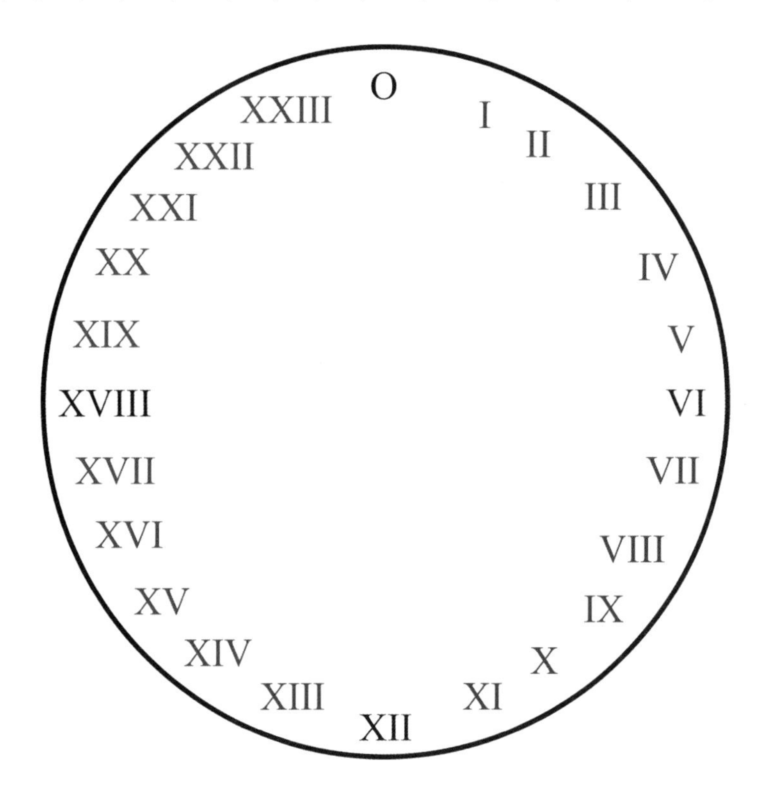

13.

1. 7:00 AM; 不合适，这个时间太早了，约翰可能在睡觉
2. 2:00 AM; 不合适，这个时间索尼娅可能在睡觉
3. 1:00 AM
4. 不是同一天，已经过了午夜，是隔天了!

14.

1. 9:35 + 4:20 = 1:55
2. 1:55 - 3:00 = 10:55 AM
3. 10:55
4. 10:10 + 1:00 = 11:10, 11:10 - 9:35 = 1小时35分钟

15.

1. (3 × 24 × 60) + (7 × 60) + 16 = 4756分钟
2. 20分钟; 1 3/4天
3. 赶不上下趟车了，距再下一趟车还要等20分钟
4. 能 (7:15 + 12 + 35 = 8:02)
5. 24 + 28 + 14 = 66天
6. 07:45; 19:45
7. XII
8. 11:00 PM

INTRODUCTION

How would you define math? It's not as easy as you might think. We know math has to do with numbers. We often think of it as a part, if not the basis, for the sciences, especially natural science, engineering, and medicine. When we think of math, most of us imagine equations and blackboards, formulas and textbooks.

But math is actually far bigger than that. Think about examples like Polykleitos, the fifth-century Greek sculptor, who used math to sculpt the "perfect" male nude. Or remember Leonardo da Vinci? He used geometry—what he called "golden rectangles," rectangles whose dimensions were visually pleasing—to create his famous *Mona Lisa*.

Math and art? Yes, exactly! Mathematics is essential to disciplines as diverse as medicine and the fine arts. Counting, calculation, measurement, and the study of shapes and the motions of physical objects: all these are woven into music and games, science and architecture. In fact, math developed out of everyday necessity, as a way to talk about the world around us. Math gives us a way to perceive the real world—and then allows us to manipulate the world in practical ways.

For example, as soon as two people come together to build something, they need a language to talk about the materials they'll be working with and the object that they would like to build. Imagine trying to build something—anything—without a ruler, without any way of telling someone else a measurement, or even without being able to communicate what the thing will look like when it's done!

The truth is: We use math every day, even when we don't realize that we are. We use it when we go shopping, when we play sports, when we look at the clock, when we travel, when we run a business, and even when we cook. Whether we realize it or not, we use it in countless other ordinary activities as well. Math is pretty much a 24/7 activity!

And yet lots of us think we hate math. We imagine math as the practice of dusty, old college professors writing out calculations endlessly. We have this idea in our heads that math has nothing to do with real life, and we tell ourselves that it's something we don't need to worry about outside of math class, out there in the real world.

But here's the reality: Math helps us do better in many areas of life. Adults who don't understand basic math applications run into lots of problems. The Federal Reserve, for example, found that people who went bankrupt had an average of one and a half times more debt than their income—in other words, if they were making $24,000 per year, they had an average debt of $36,000. There's a basic subtraction problem there that should have told them they were in trouble long before they had to file for bankruptcy!

As an adult, your career—whatever it is—will depend in part on your ability to calculate mathematically. Without math skills, you won't be able to become a scientist or a nurse, an engineer or a computer specialist. You won't be able to get a business degree—or work as a waitress, a construction worker, or at a checkout counter.

Every kind of sport requires math too. From scoring to strategy, you need to understand math—so whether you want to watch a football game on television or become a first-class athlete yourself, math skills will improve your experience.

And then there's the world of computers. All businesses today—from farmers to factories, from restaurants to hair salons—have at least one computer. Gigabytes, data, spreadsheets, and programming all require math comprehension. Sure, there are a lot of automated math functions you can use on your computer, but you need to be able to understand how to use them, and you need to be able to understand the results.

This kind of math is a skill we realize we need only when we are in a situation where we are required to do a quick calculation. Then we sometimes end up scratching our heads, not quite sure how to apply the math we learned in school to the real-life scenario. The books in this series will give you practice applying math to real-life situations, so that you can be ahead of the game. They'll get you started—but to learn more, you'll have to pay attention in math class and do your homework. There's no way around that.

But for the rest of your life—pretty much 24/7—you'll be glad you did!

1 TIME CONVERSION

Sonia has been thinking about time a lot lately. She just got a cell phone, so she can check the time whenever she wants. Every time she gets her phone out, she looks at the time. And every day she frequently looks at the clock in class, to see how much longer is left until school is over. It's almost summer break, and she's ready to play sports, go on vacation, and hang out with her family and friends.

Sonia has a countdown till summer break that's running on her computer at home. It counts down in both days, hours, and minutes, and also in just minutes. Looking at just the minutes makes it seem like summer is really long away, but looking at the days makes it shorter. Check out the next page to see how Sonia's countdown converts minutes to days and hours.

An hour is made up of 60 minutes, and a minute is made up of 60 seconds—easy enough.

But what happens when you want to figure out how many minutes are in 5 hours, or 9 hours and 23 minutes. What if you wanted to find out how many seconds made up that time?

It's pretty simple to **convert** between hours and minutes (or minutes and seconds). If you know the number of hours, just multiply it by 60.

1. How many minutes are in 3 hours?

   If there are extra minutes in addition to the hours, just multiply the hours and add on the minutes to your answer.

2. How many minutes are in 3 hours and 56 minutes?

   To find how many hours are contained in a certain number of minutes, divide by 60.

3. How many hours are in 480 minutes?

The tricky part is when you have a remainder when you divide the hours by 60. Let's say you are looking at 515 minutes. When you divide 515 by 60, you get 8.58. How many minutes are in .58 hours?

Once you have divided the minutes by 60 and seen there is a remainder, next figure out how many minutes are in the number of hours in the ones' place. Then subtract that from the number of total minutes. What is left is the number of extra minutes past the hour:

515 minutes in 8.58 hours – 480 minutes in 8 hours = 35 minutes
So, 515 minutes = 8 hours 35 minutes

4. How many hours and extra minutes are in 171 minutes?

5. If Sonia has 2 days, 7 hours, and 46 minutes left until summer break, how many minutes is that? Remember, there are 24 hours in a day.

2 TIME FRACTIONS

Sonia is talking to her friends about their after-school plans that day. She wants to go hang out in the park, but she has to meet with a teacher right after school.

"It'll only take half an hour," she says.

Her friend Jackson points out that the teacher she's meeting with really likes to talk. He says, "I think it will be more like three-quarters of an hour."

Sonia agrees and suggests they all meet up at the park at a quarter past four, just to be sure she'll have enough time to meet with her teacher and then get to the park.

Sonia and her friends are talking about time in terms of fractions. Just like with anything else, parts of an hour can be **expressed** in fractions. Do some practice on the next page.

Most people memorize the common fractions we use for time:

one-quarter (¼) hour = 15 minutes
half (½) an hour = 30 minutes
three-quarters (¾) of an hour = 45 minutes

1. What time are Sonia and her friends meeting if they plan to arrive at quarter after 4?

You can also figure out how many minutes are in fractions of an hour using math. All you need to do is divide the number of minutes in an hour by the denominator. Then multiply that number by the numerator.

To find three-quarters of an hour:

¾ of an hour = (60 minutes ÷ 4) x 3
¾ of an hour = 45 minutes

2. What is two-quarters of an hour using math? What else could you say instead of two-quarters of an hour?

You can use this method to come up with fractions of time we don't commonly use, like a fifth of an hour or two-thirds of an hour.

3. What is one-fifth of an hour?

4. What is two-thirds of an hour?

Hours aren't the only unit of time you can divide into fractions. You can also talk about minutes, days, weeks, months, years, and even centuries in fractions. Try finding the following fractions of time:

5. three-quarters of a week (in days) =
one-half of a minute (in seconds) =
four-fifths of a 30-day month (in days) =
seven-twelfths of a year (in months) =
one-quarter of a century (in years) =

3 FUTURE TIME

After school, Sonia and her friends meet up at quarter after four and hang out in the park. After a little while, Jackson mentions to everyone that he has to leave in a couple hours to go visit his grandparents. Sonia and Megan also need to go home to finish a little bit of studying they have left before the end of the school year. They have a test tomorrow, and they want to make sure they do well.

How much time do the friends have left to hang out? How much time will they need to study and sleep? And how much time is left before Sonia and Megan's test? Figure it out with time addition.

Sonia and her friends start talking about time at 4:35. Jackson will have to leave in an hour and a half to get home in time to visit his grandparents. What time will Jackson have to leave? You can do addition with time just like with any other number.

First, you will need to convert an hour and a half into time. An hour and a half is like saying 1:30. So:

$$4:35 + 1:30 = 5:65$$

Since there are only 60 minutes in an hour, you need to carry the extra 5 minutes into another hour. You can express the 60 as 00, the beginning of a new hour, so 5:65 becomes 6:05.

1. It takes Jackson about 40 minutes to get to his grandparents' house. What time will he get there if he and his family leave his house at 6:30?

Sonia and Megan both want to be home by 6:30 to have dinner and do some studying. Megan lives right next to the park, so she can leave at 6:30. Sonia lives a little further away than Megan, so she has to leave 10 minutes earlier. How much longer do both girls have to hang out at the park?

These problems need subtraction. You know what time they need to be home, but you don't know how much time they have until then.

Megan will leave at 6:30, and it is 4:35 now. Set up the subtraction problem like usual. But remember, there are only 60 minutes in an hour.

$$6:30 - 4:35 = 1:55$$

In other words, Megan has an hour and 55 minutes before she needs to leave the park.

To find how much longer Sonia has at the park, you can do another subtraction problem using 6:20 (10 minutes earlier than Megan), or you can just subtract 10 minutes from the time Megan has left at the park.

2. How much time is there before Sonia has to leave if it is 4:35 now?

4
PAST TIME

When Sonia goes home, she eats dinner and starts to study for her math test, which is tomorrow. She thinks about all the time she has spent studying for this test, and for other math tests throughout the year. She wonders just how much time she has spent on math! She wants to add it all up, but figuring how much time she has spent studying will also take some subtraction. See how Sonia does it on the next page.

First, Sonia makes a list of all the time she has spent studying for the last few tests:

Test #1
Monday: half an hour
Tuesday: one hour
Wednesday: two hours and 15 minutes

Test #2:
Saturday: 45 minutes
Sunday: two hours
Monday: one hour and 20 minutes

Test #3
Thursday: 4:00–5:30
Saturday: 1:15–3:00
Sunday: 7:00–7:30
Monday: 7:00–9:15

1. How many hours did Sonia spend studying for each test?

2. What are those three amounts in minute-only form?

Sonia can't remember how much she studied for the tests before these three. She does know she has had 15 tests in math throughout the year.

Even though she doesn't know the exact number of hours she studied for all her tests, she can find the average number of hours she studied for these three tests. Once she finds the average, she can multiply it by the number of tests she has taken to approximate just how many hours she has spent studying in total.

First, find the average number of hours she studied for these three. Add all the hours together and divide by the number of tests for which you counted hours. It's easier to find the average in just minutes (rather than minutes and hours).

3. What is the average number of minutes she studied?

Now multiply the average number of minutes she studied per test by the number of math tests she has taken all year. Your calculation will give you an estimate of how many minutes she studied all year for math.

4. How many minutes did she study? What is it in hours and minutes?

5 TIME PATTERNS

Sonia usually takes the school bus to school, but once in a while she misses it. If that happens, there is also a public bus that stops by her corner. She can take that bus and transfer to another bus that will take her to school.

She doesn't like to miss the school bus, because taking the public bus is longer and more hassle. However, this morning, she misses her bus. On top of that, it's the last day of school! She wants to be there to say goodbye to her friends and teachers, and get some of her assignments back.

Sonia checks the public bus schedule. She has to look at two different schedules and line them up because she has to take two different buses. The buses come at regular intervals, so

Sonia has to understand time patterns in order to get to school on time. Can she do it?

The school bus normally comes around 7:25. Sonia just missed it, and it is now 7:28. She checks the bus schedule times at the stops she needs and sees:

Bus #1	**Bus #2**
7:18	7:15
7:28	7:29
7:38	7:45
7:48	7:59
7:58	8:15
8:08	8:29

1. How many minutes are between each Bus #1 arrival?

2. What is the pattern of Bus #2 arrivals in minutes?

The first bus ride takes about 9 minutes. Then she will get off and wait for the next bus. The second bus ride to school takes about 7 minutes.

Sonia has already missed the 7:18 bus and just missed the 7:28 Bus #1. She will have to take the 7:38.

3. What time will she get to the second bus stop if she gets on the first bus at 7:38? Will she make it to the 7:45 Bus #2?

4. Which Bus #2 will Sonia need to take?

Sonia's school day starts at 8:10. She normally gets there at 7:50 on her school bus, and has a few minutes to relax before school starts.

5. Will Sonia get to school on time considering the time she gets picked up by Bus #2? What time will she arrive?

6. How many minutes later is she than when she would normally get to school?

6
BEING ON TIME

Sonia is tired of missing the bus and being late for other things. It seems like she's always running late! This summer, she wants to practice being on time. When the next school year arrives, she will always be on time!

On the first day of her summer vacation, Sonia has a dentist appointment. She isn't very happy about it, but she decides the appointment is a good chance to practice being on time. Her dad, who is going to take her to her appointment, also tends to be late and has trouble getting places on time. Sonia thinks she'll have to be the one to get them there on time. The next pages show some calculations she could do to help her be on time.

Sonia's dentist appointment is at 3:30 in the afternoon. She knows it takes 20 minutes to get there.

1. What time do you think Sonia should leave to get to her dentist appointment?

There are some other things to consider. She and her dad also have to fill out some paperwork at the dentist. She's not sure how long that will take, but she thinks maybe 10 minutes. Also, there is some construction along the way to the dentist's office. Sonia is also not sure how much longer it will take to get there, but she takes a guess of 15 minutes extra.

2. How many minutes do they need to add to get to the dentist on time? What time should they leave?

Sonia almost forgets about her appointment, but she looks at her phone at 2:40 and remembers. She has to remind her dad and get ready to go. They are out the door by 2:46.

3. If the trip ends up taking 33 minutes, will they get there on time for the appointment, including filling out paperwork? If so, how much extra time do they have? If not, how late

were they?

From now on, Sonia will use a chart that looks like this. She has soccer camp next week, which starts at 9:00 AM on Monday, Wednesday, and Friday. On Tuesday and Thursday, it starts at 10:00 AM. It will take her 12 minutes to get there. Add those lines to the chart, as well as the information for the dentist appointment.

Event	**Day**	**Time**	**How Long It Takes to Get There**	**Time Needed to Leave**
Dentist appointment	Wednesday	3:30		
Soccer camp				
Soccer camp				

7 ALARMS

As part of Sonia's plan to be on time, she wants to use alarms more often. During the school year, she uses an alarm clock to wake up every day for school, but she sometimes sleeps past the alarm.

Sonia has an alarm on her new phone. She thinks she should use it to set alarms so she's on time for things. She knows she might not always pay attention to just one alarm, though, so she'll need to set multiple alarms to remind her to do things. See how she does it on the next pages.

One day next week, Sonia will have to use her alarm clock to get up. She's going to an amusement park with her friend Marco. He and his family are picking her up at 8:30 in the morning, so she has to be up and ready to go by then.

Sonia needs at least 45 minutes to get dressed, eat breakfast, and brush her teeth.

1. What is the earliest she can get up?

She isn't sure she'll be able to get up with just one alarm, though. Her alarm clock has a snooze button. When her alarm goes off, she can press the snooze button, which will give her 7 more minutes of sleep. The alarm goes off after 7 minutes.

Write down how many minutes she adds to her sleep by pressing snooze, and what time Sonia will get up. The first one has been done for you.

2. 0 snooze = 0 minutes, 7:45
   1 snooze =
   2 snooze =
   3 snooze =
   4 snooze =

Sonia thinks she may hit the snooze button twice.

3. What time should she set the alarm for if she hits snooze twice?

On the day of the trip, Sonia actually ends up hitting the snooze button 3 times and has to rush to get ready.

4. What time did she get up if she set the alarm thinking she would only hit snooze twice? How many minutes does she have to get ready?

8 STOPWATCHES

At the amusement park, Sonia and Marco have a great time. They ride roller coasters, play games, and eat pizza.

Then they find the go-carts. The amusement park has a go-cart setup that allows customers to race. Every day they put up the fastest times for people to beat later in the day. Sonia and Marco watch for a while before they take a turn themselves.

After a few minutes, they notice the times being posted for the races don't seem to match the real times. Sonia uses the stopwatch on her phone to time how long it really takes the racers to finish. Once they figure it out, they're ready to race too—and win!

The racers have to drive around the track 5 times. The top scores for the day are:

3 minutes 3 seconds
3 minutes 17 seconds
3 minutes 24 seconds
3 minutes 39 seconds
3 minutes 41 seconds

1. What is the difference between first place and fifth place?

Once Sonia and Marco notice the times on the board don't seem right, they check with their own stopwatch. The person who finishes first in this race gets a time of 3 minutes 3 seconds, so he move to first place on the board. However, the time they get is 4 minutes 53 seconds.

2. By how much is the amusement park stopwatch off?

To be sure, they time another race. This time, the winner of the race gets 3 minutes 38 seconds on the board. They timed her getting 5 minutes 28 seconds.

3. Is the amusement park stopwatch off by the same amount? If not, how much was it off this time?

Marco is going to race and see if he can beat the best score for the day. Sonia agrees to stand on the sidelines and keep an eye on the stopwatch. When it's getting close to the actual time he needs to win, she'll yell so he knows he's running out of time.

As Marco is about to drive his final lap, Sonia yells that the stopwatch says 2 minutes and 30 seconds.

4. What's the average amount of time it has taken him to make 1 lap?

5. Will he beat the best score if his last lap takes the average amount of time the other laps have? What will his time be? If not, how far away from the top score will he be?

9 CALENDAR MATH

Sonia is getting better and better at being on time. She still tends to forget about places she has to be that are days away, though, and she's often late because she only remembers them at the last minute. For example, she forgot her friend Megan's birthday last week until another friend called her and asked her if she was going, just a half hour before the party started.

Looking at the calendar and using a planner might help Sonia. She can remind herself ahead of time that she needs to be somewhere. And she can use math to use her calendar, and remember important dates!

The next big thing Sonia has to remember is her dad's birthday, which is on July 31. She wants to get him a card and a present, and she doesn't want to forget.

1. How many days away is her dad's birthday if today is July 26?

She can write it in her calendar and then remind herself when her dad's birthday is 2 days away.

2. How many days away does she have until she needs to remind herself about the birthday?

Sonia also wants to make the most of her summer. She isn't sure how many days are left of summer break, but having a calendar makes it easy.

She can choose from a few ways to figure out how much time she has left during summer break. She could count out every single day on the calendar, but that will take a lot of time, and she might end up miscounting.

She could count by week instead. That's a lot easier than counting days. Then she can just multiply the number of weeks by 7 to figure out just how many days she has left.

3. If there are 5 weeks and 5 days until she goes back to school, how many days of summer does she have left?

Sonia could also add up the number of days in the months she has left until school starts. Today is July 26. School starts on September 4.

To add correctly, you need to know how many days there are in each month. Some have 30 and other have 31 (and one has 28).

28 days: February
30 days: April, June, September, November
31 days: January, March, May, July, August, October, December

4. Using this information, how many days does Sonia have until school starts?

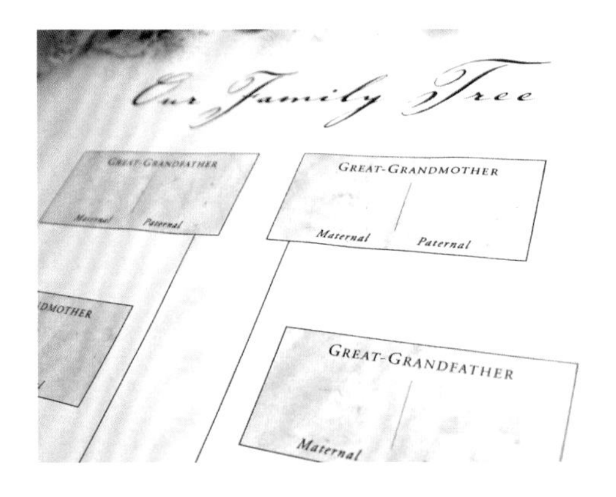

10 TIMELINES

Sonia's mom is putting together a family history. She is going through lots of old photos and letters, trying to figure out who is related to whom. Sonia helps out sometimes, because she thinks her family history is interesting. She finds it hard to keep track of everything, though, because there are so many documents and so many dates.

She and her mom start putting together a timeline for their family. Every time they find a new piece of information, like when someone was born, they add it to the timeline. When they are finished, they will have a long timeline that lists the important things that have happened in the family. Check out the next page to see what a timeline looks like, and help them construct their family timeline.

Here are some of the pieces of information Sonia and her mom have discovered from 1900 to 1960:

Great-grandmother born: September 30, 1908
Great-grandfather born: March 4, 1912
Great-grandparents on father's side came to America: 1927
Grandmother born: 1941
Great-grandfather died: 1953
Mother's family came to America: 1958
Grandfather on father's side started business: 1960

Timelines are arranged from oldest dates on the left to newest dates on the right. The dates are shown **chronologically**. Sonia's timeline lists specific days with years.

Arrange the rest of Sonia's family's information on the timeline below. The information above is written down as they found it, so it's not chronological yet. The information can go either above or below the line, so that it will all fit.

11 MILITARY TIME

Sonia goes over to her friend Natalia's house one afternoon. They play a computer game for a while. At one point, Sonia notices the clock on the computer reads 16:34. "What does that mean?" Sonia asks, pointing to the clock.

Natalia explains the computer is showing military time, or 24-hour time as it's also known. (In the United States, the military uses a 24-hour clock, but everyone else does not.) Natalia's parents are from Russia, where people tell time a different way from what they do in North America. They keep the clock on their computer in 24-hour format because that's what they're used to.

Sonia looks at her own clock and sees that it is 4:34. She thinks she can see the connection between the way she tells time and military time. Can you?

In this part of the world, we call the way we tell time the 12-hour clock. In 12-hour time, the same time number happens twice in one day. For example, 8:04 happens once in the morning and once in the evening. The times start at midnight and start over at noon. We use AM and PM to show which time we are talking about.

In military time, no time is ever repeated in a day. Instead, every hour has its own number, from 1 through 24. After 12:00 PM comes 13:00, not 1:00 again. AM and PM aren't necessary. Times between 1:00 AM to 12:59 PM look exactly the same in 12-hour time and military time. 11:30 AM in 12-hour time is 11:30 in military time. Times with numbers only in the ones' places sometimes have a zero in front of them, so 2:00 AM would be 02:00 in military time.

1. What is 3:54 AM in military time?

Where military time is different is between 1:00 PM and 12:59 AM.

The way to figure out a time like 16:34 is to subtract 12:00 from it. The remainder is the time in 12-hour clock format. So:

$$16:34 - 12:00 = 4:34$$

The minutes will always be the same between the two ways of telling time. Only the hours might be different.

2. What time is 19:20 in 12-hour time?

3. How about 22:45?

Do you have a guess what 12:00 AM is in military time? People start to count the day at 12:00 AM. They start at zero, so midnight is 00:00!

4. Convert between the two below. Don't forget the AM or PM.

   6:00 PM = __________
   ____________ = 05:30
   ____________ = 23:18
   12:09 PM = _________
   ____________ = 00:20

12 ROMAN NUMERALS

While Sonia is at Natalia's house, she also notices a clock that has some strange-looking numbers on it. Instead of 1 through 12, it has letters like I and X. Again, she has to ask Natalia how to tell time in her house!

Natalia says the clock is in Roman **numerals**. Long ago, people in the Roman empire used to use Roman numerals all the time. Today, people use different numerals (1, 2, 3, etc.). One of the places Roman numerals are still used today, though, is on clocks. The years of big sports events like the Super Bowl and the Olympics are also often written in Roman numerals. It's worth figuring out Roman numerals in case you ever have to read them.

Clocks are great places to learn how to read Roman numerals, because you already know that a clock lists numbers 1 through 12.

Here are some of the most common Roman numerals:

I = 1
V = 5
X = 10
L = 50
C = 100
M = 1,000

On a clock, you only have to deal with I, V, and X.

You form numbers by adding different letters together. Two would be two 1s, or II. Three would be three 1s, or III. But four is a little different. You don't want to list out four IIIIs in a row. Instead, you write IV, which means one less than five. You do the same thing for nine, but using an X instead of a V.

Here is 1 through 6 in Roman numerals: I, II, III, IV, V, VI

1. What are 7 through 12, to finish out the clock?

Natalia also happens to have a 24-hour clock written in Roman numerals. Instead of 12 hours, there are 24 on the clock face.

2. Fill in the rest of the Roman numerals on the clock below:

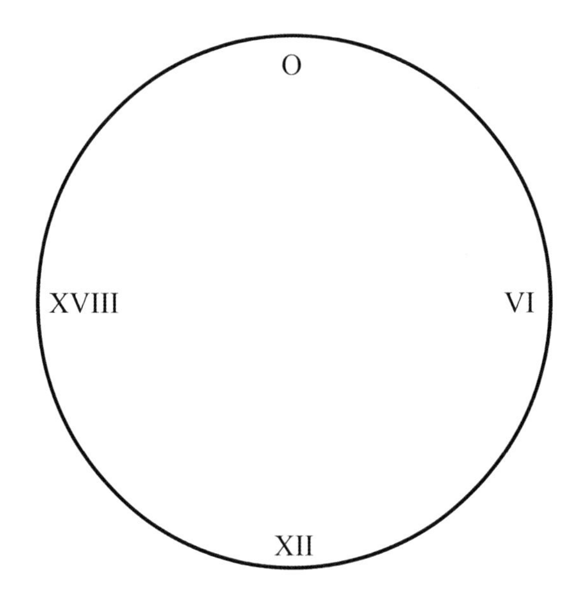

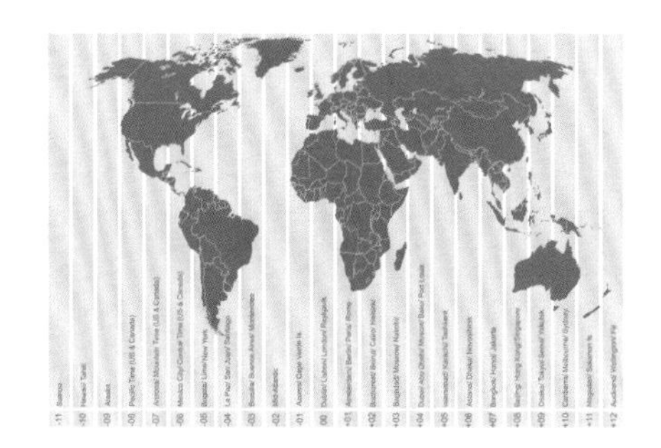

13 TIME ZONES

One of Sonia's friends from school is moving this summer. Johan's mom got a new job in California, so he and his family are moving. Sonia is really sad that he's moving, but she promises to keep in touch. The problem is that California is so far away. Sonia lives on the East Coast, which is thousands of miles from California. The two places are also in different time zones. When it is one time for Sonia, it will be earlier in California.

How can time be different? Time is just how we measure the movement of the sun (or really, the measure of how fast the Earth goes around the sun). In general, we say that it is noon when the sun is at its highest point in the sky. But at different places on the Earth, the sun will be at its highest point at different times. Right now, it might be noon where you live. But on the opposite side of the world, it will be really dark. Instead of saying that it's noon over there, we adjust time so that it will be noon when the sun is high in the sky.

We end up creating time zones. As you move around the Earth from east to west, every few hundred miles, it gets an hour earlier. So if it is 3:00 PM where you are now, a few hundred miles to your west it is an hour earlier. And a few hundred miles to your east, it is an hour later. Time doesn't change when you go north or south, though.

Sonia will have to think about this when she keeps in touch with Johan. She doesn't want to end up calling him when he's sleeping, and she doesn't want him calling while she's sleeping either. She will have to figure out the best times to call him.

The time zone on the East Coast is called the Eastern Time Zone. The time zone in California is called the Pacific Time Zone. There are three hours of difference between them.

When it is 4:00 PM where Sonia lives, it will be 3 hours earlier in California. Subtract when you are trying to find out what time it is in an earlier time zone:

$$4{:}00 - 3{:}00 = 1{:}00 \text{ PM}$$

1. What time would it be in California if it is 10:00 AM in Sonia's town? Is that a good time to call Johan?

Johan will also have to know how to figure out the time on the East Coast. He wants to call his friends, and some of his family will still be there.

When it is 4:00 PM where Johan is moving, it will be 3 hours later in his old town. To find out what time it is in a later time zone, add the time:

4:00 PM + 3:00 = 7:00 PM

2. What time is it for Sonia when it is 11:00 PM in Johan's new home? Is that a good time for him to call? Why or why not?

The bigger the distance, the bigger the time difference is. For Natalia, whose grandparents live outside of Moscow in Russia, the time difference is really big. Right now, her family is 8 hours ahead of her.

3. If it is 5:00 PM for Natalia, what time is it for her grandparents?

4. Is it still the same day for Natalia's grandparents as it is for her? Why or why not?

14 TIME ZONE TRAVEL

Some day, Sonia wants to go visit Johan in California. She has never been to California, and she wants to go on a vacation. She also wants to see Johan again!

Traveling to California would involve time zone changes. Instead of just talking to someone in a different time zone, Sonia would actually be in a different time zone!

Changing time zones can get confusing, and can mess up your **sleep schedule**. It's not too hard to figure out once you know the math. Try it out on the next page.
You already know that California is 3 hours earlier than where Sonia lives. If she were to visit Johan, she would be gaining hours in her day.

This is a flight schedule Sonia might have if she went to California:

Departure
Leave: 9:35 AM
Flight length: 4 hours 20 minutes

1. If she were traveling in the same time zone, what time would she get to California?

However, she is actually going to arrive in California 3 hours earlier than she would have if there were no time changes. Just subtract 3 hours from the time she would have arrived.

2. What time in California is it when Sonia arrives?

Another way to figure out what time it is when she arrives is to find the difference between the flight time and the time zone change. Then add the difference to the time she is leaving from the East Coast:

3. 4 hours 20 minutes – 3 hours = 1 hour 20 minutes

$$1{:}20 + 9{:}35 =$$

Sonia might also have to change planes on the way to California. She might have to stop in Chicago, which is one hour behind her home time zone.

If her flight landed in Chicago at 10:10 in Chicago time, how long was the flight from the airport she started at?

Change the Chicago time into time on the East Coast. Then subtract the time she left from your answer.

4. How long was the flight?

15 PUTTING IT ALL TOGETHER

Sonia has learned a lot about time over the summer. She has figured out how to be on time, she can read Roman numerals, and she knows how to tell military time. See if you can remember some of what she has learned during her summer break.

1. How many minutes are in 3 days, 7 hours, and 16 minutes?

2. How many minutes are in one-third of an hour?

   And how many days are in one-quarter of a week?

3. You need to catch a bus at 3:07. The next bus is at 3:29. Right now it is 2:58 and it takes you 11 minutes to walk to the bus stop.

   Will you catch the bus?

   If not, how long will you have to wait until the next one?

4. Your snooze button is set for 6 minutes. You set your alarm for 7:15 AM and end up hitting the snooze button twice.

   If it takes you 35 minutes to get ready and walk to school, will you make it by the time school starts at 8:05?

5. Today is January 7. You need to remember you have a school concert on March 14.

   How many days away is that?

6. What is 7:45 AM in military time?

What is 7:45 PM in military time?

7. What is the Roman numeral form of twelve?

8. You friend lives two time zones ahead of you (to the east).

   What time is it in your friend's time zone when it is 9:00 PM where you live?

Answers

1.

1. 3 x 60 = 180 minutes
2. 180 + 56 = 236 minutes
3. 480/60 = 8 hours
4. 2 hours 51 minutes
5. (24 x 2 x 60) + (7 x 60) + 46 = 3,346 minutes

2.

1. 4:15
2. 30 minutes; half an hour
3. 12 minutes
4. 40 minutes
5. 5 ¼ days
   30 seconds
   24 days
   7 months
   25 years

3.

1. 6:30 + :40 = 7:10
2. 1 hour and 45 minutes

4.

1. Test #1 = 3 hours 45 minutes; Test #2 = 4 hours 5 minutes; Test #3 = 6 hours
2. 225 minutes, 245 minutes, 360 minutes
3. 276.67 minutes
4. 4150 minutes; 69 hours and 10 minutes.

5.

1. 10
2. Every 16 minutes, alternating with every 14 minutes.
3. 7:47; no
4. The 7:59
5. Yes, she will get there at 8:06.
6. 16 minutes later

6.

1. 3:10
2. 25 extra minutes; 2:45
3. Yes, they got there on time and had 11 minutes before her appointment.

Event	**Day**	**Time**	**How Long It Takes to Get There**	**Time Needed to Leave**
Dentist appointment	Wednesday	3:30	45 minutes	2:45
Soccer camp	Monday, Wednesday, Friday	9:00	12 minutes	8:48
Soccer camp	Tuesday, Thursday	10:00	12 minutes	9:48

7.

1. 7:45
2. 7 minutes, 7:52
   14 minutes, 7:59

21 minutes, 8:06
28 minutes, 8:13
3. 7:31
4. 7:31 + :21 = 7:52; 8:30 – 7:52 = 38 minutes to get ready

1. 35 seconds
2. 4:53 – 3:03 = 1 minute 50 seconds
3. Yes, it is still off by 1 minute 50 seconds.
4. 37.5 seconds
5. Not quite; 3 minutes and 7.5 seconds (150 seconds + 37.5 seconds = 187.5 seconds = 3 minutes and 7.5 seconds); he will be 4.5 seconds away from beating the top score.

9.

1. 5 days
2. 3 days
3. 40 days
4. 40 days

10.

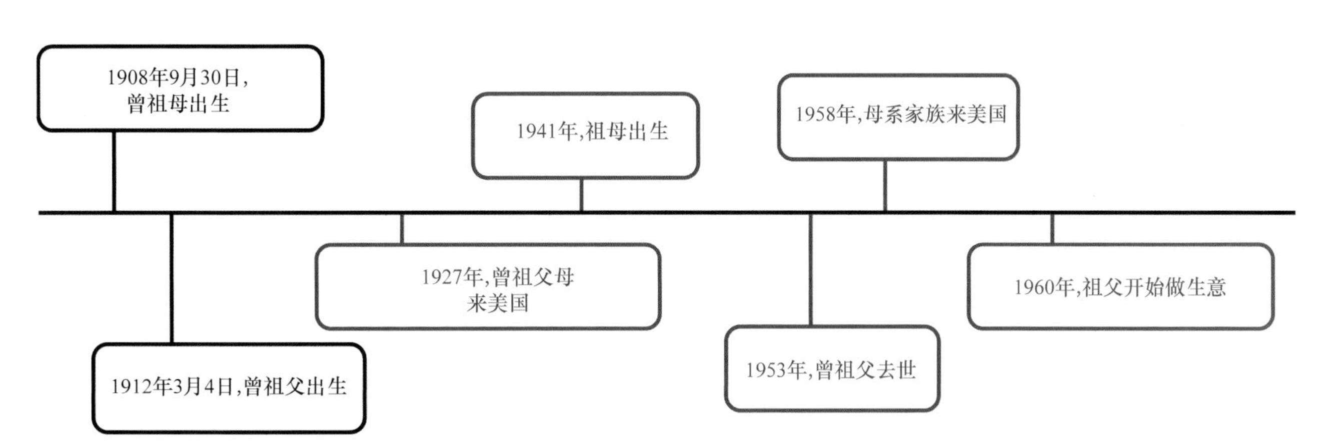

11.

1. 03:54
2. 7:20 PM
3. 10:45PM
4. 18:00
   5:30 AM
   11:18 PM
   12:09
   12:20 AM

12.

1. VII, VIII, IX, X, XI, XII
2. I, II, III, IV, V, VII, VIII, IX, X, XI, XIII, XIV, XV, XVI, XVII, XIX, XX, XXI, XXII, XXIII

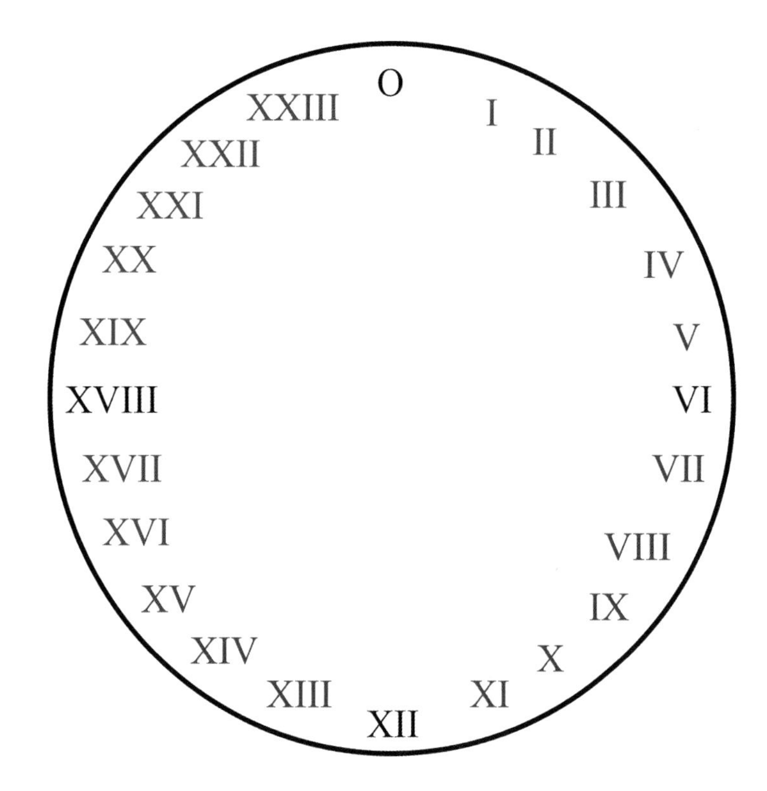

13.

1. 7:00 AM; no, it is probably too early and Johan will be sleeping.
2. 2:00 AM; no, it is not a good time because she will be sleeping.
3. 1:00 AM

4. No, it is past midnight, so it is the next day there!

14.

1. 9:35 + 4:20 = 1:55
2. 1:55 – 3:00 = 10:55 AM
3. 10:55
4. 10:10 + 1:00 = 11:10, 11:10 – 9:35 = 1 hour 35 minutes

15.

1. (3 x 24 x 60) + (7 x 60) + 16 = 4756 minutes
2. 20 minutes; 1¾ days
3. No, you'll miss it. You'll have to wait 20 minutes from the time you get to the bus stop.
4. Yes (7:15 + 12 + 35 = 8:02)
5. 24 + 28 + 14 = 66 days
6. 07:45; 19:45
7. XII
8. 11:00 PM